Weight-of-evidence for Forensic DNA Profiles

Statistics in Practice

Founding Editor

Vic Barnett
Nottingham Trent University, UK

Statistics in Practice is an important international series of texts, which provide detailed coverage of statistical concepts, methods and worked case studies in specific fields of investigation and study.

With sound motivation and many worked practical examples, the books show in down-to-earth terms how to select and use an appropriate range of statistical techniques in a particular practical field within each title's special topic area.

The books provide statistical support for professionals and research workers across a range of employment fields and research environments. The subject areas covered include medicine and pharmaceutics; industry, finance and commerce; public services; the earth and environmental sciences, and so on.

The books also provide support to students studying statistical courses applied to the above areas. The demand for graduates to be equipped for the work environment has led to such courses becoming increasingly prevalent at universities and colleges.

It is our aim to present judiciously chosen and well-written workbooks to meet everyday practical needs. The feedback of views from readers will be most valuable to monitor the success of this aim.

A complete list of titles in this series appears at the end of the volume.

Weight-of-evidence for Forensic DNA Profiles

David J. Balding

Professor of Statistical Genetics
Imperial College,
London, UK

John Wiley & Sons, Ltd

Copyright 2005   John Wiley & Sons Ltd, The Atrium, Southern Gate, Chichester,
                       West Sussex PO19 8SQ, England

                       Telephone (+44) 1243 779777

Email (for orders and customer service enquiries): cs-books@wiley.co.uk
Visit our Home Page on www.wileyeurope.com or www.wiley.com

All Rights Reserved. No part of this publication may be reproduced, stored in a retrieval system or
transmitted in any form or by any means, electronic, mechanical, photocopying, recording, scanning
or otherwise, except under the terms of the Copyright, Designs and Patents Act 1988 or under the
terms of a licence issued by the Copyright Licensing Agency Ltd, 90 Tottenham Court Road, London
W1T 4LP, UK, without the permission in writing of the Publisher. Requests to the Publisher should
be addressed to the Permissions Department, John Wiley & Sons Ltd, The Atrium, Southern Gate,
Chichester, West Sussex PO19 8SQ, England, or emailed to permreq@wiley.co.uk, or faxed to (+44)
1243 770620.

This publication is designed to provide accurate and authoritative information in regard to the subject
matter covered. It is sold on the understanding that the Publisher is not engaged in rendering
professional services. If professional advice or other expert assistance is required, the services of a
competent professional should be sought.

Other Wiley Editorial Offices

John Wiley & Sons Inc., 111 River Street, Hoboken, NJ 07030, USA

Jossey-Bass, 989 Market Street, San Francisco, CA 94103-1741, USA

Wiley-VCH Verlag GmbH, Boschstr. 12, D-69469 Weinheim, Germany

John Wiley & Sons Australia Ltd, 33 Park Road, Milton, Queensland 4064, Australia

John Wiley & Sons (Asia) Pte Ltd, 2 Clementi Loop #02-01, Jin Xing Distripark, Singapore 129809

John Wiley & Sons Canada Ltd, 22 Worcester Road, Etobicoke, Ontario, Canada M9W 1L1

Wiley also publishes its books in a variety of electronic formats. Some content that appears
in print may not be available in electronic books.

Library of Congress Cataloging-in-Publication Data

Weight-of-evidence for forensic DNA profiles / David J. Balding.
    p. cm. – (Statistics in practice)
  Includes bibliographical references and index.
  ISBN 0-470-86764-7 (alk. paper)
    1. Forensic genetics – Statistical methods. I. Balding, D. J. II. Series.

RA1057.5.A875 2005
614′.1′0727 – dc22
                                                    2004061587

British Library Cataloguing in Publication Data

A catalogue record for this book is available from the British Library

ISBN 0-470-86764-7

Produced from LaTeX files supplied by the authors and processed by Laserwords Private Limited,
Chennai, India

To Emily Laura

Contents

Sections marked † are at a higher level of specialization and/or are peripheral topics not required to follow the remainder of the book. They may be skipped at first reading.

Preface — xi

1 Introduction — 1
  1.1 Weight-of-evidence theory — 1
  1.2 About the book — 3
  1.3 DNA profiling technology — 3
  1.4 What you need to know already — 4
  1.5 Other resources — 5

2 Crime on an island — 7
  2.1 Warm-up examples — 7
    2.1.1 Disease testing: Positive Predictive Value (PPV) — 7
    2.1.2 Coloured taxis — 9
  2.2 Rare trait identification evidence — 10
    2.2.1 The "island" problem — 10
    2.2.2 A first lesson from the island problem — 11
  2.3 Making the island problem more realistic — 13
    2.3.1 Uncertainty about p — 14
    2.3.2 Uncertainty about N — 15
    2.3.3 Possible typing errors — 15
    2.3.4 Searches — 17
    2.3.5 Other evidence — 18
    2.3.6 Relatives and population subdivision — 19
  2.4 Weight-of-evidence exercises — 20

3 Assessing evidence via likelihood ratios — 22
  3.1 Likelihood ratios — 22
  3.2 The weight-of-evidence formula — 24
    3.2.1 Application to the island problem — 25
    3.2.2 The population $\mathcal{P}$ — 25

3.3	General application of the formula	27
	3.3.1	Several items of evidence	27
	3.3.2	Assessing all the evidence	29
	3.3.3	The role of the expert witness	30
3.4	Consequences for DNA evidence	31
	3.4.1	Many possible culprits	31
	3.4.2	Incorporating the non-DNA evidence	31
	3.4.3	Relatives	33
	3.4.4	Laboratory and handling errors	34
	3.4.5	Database searches	35
3.5	Some derivations †	36
	3.5.1	Bayes theorem for identification evidence	37
	3.5.2	Uncertainty about p and N	38
	3.5.3	Grouping the alternative possible culprits	39
	3.5.4	Typing errors	40
3.6	Further weight-of-evidence exercises	40

4 Typing technologies — 43
4.1	STR typing	44
	4.1.1 Anomalies	46
	4.1.2 Contamination	49
	4.1.3 Low copy number (LCN) profiling	50
4.2	mtDNA typing	50
4.3	Y-chromosome markers	51
4.4	X-chromosome markers †	52
4.5	SNP profiles	53
4.6	Fingerprints †	54

5 Some population genetics for DNA evidence — 56
5.1	A brief overview	56
	5.1.1 Drift	56
	5.1.2 Mutation	59
	5.1.3 Migration	60
	5.1.4 Selection	60
5.2	θ, or F_{ST}	62
5.3	A statistical model and sampling formula	63
	5.3.1 Diallelic loci	63
	5.3.2 Multi-allelic loci	68
5.4	Hardy–Weinberg equilibrium	69
	5.4.1 Testing for deviations from HWE †	70
	5.4.2 Interpretation of test results	74
5.5	Linkage equilibrium	75
5.6	Coancestry †	77
5.7	Likelihood-based estimation of θ †	79
5.8	Population genetics exercises	81

CONTENTS ix

6 Identification 82
6.1 Choosing the hypotheses 82
6.1.1 Post-data equivalence of hypotheses 84
6.2 Calculating likelihood ratios 85
6.2.1 The match probability 85
6.2.2 One locus 87
6.2.3 Multiple loci: the "product rule" 89
6.2.4 Relatives of s 90
6.2.5 Confidence limits † 92
6.2.6 Other profiled individuals 93
6.3 Application to STR profiles 94
6.3.1 Values for the p_j 95
6.3.2 The value of θ 96
6.3.3 Errors 98
6.4 Application to haploid profiles 99
6.4.1 mtDNA profiles 99
6.4.2 Y-chromosome markers 101
6.5 Mixtures 101
6.5.1 Visual interpretation of mixed profiles 101
6.5.2 Likelihood ratios under qualitative interpretation 103
6.5.3 Quantitative interpretation of mixtures 108
6.6 Identification exercises 109

7 Relatedness 111
7.1 Paternity 111
7.1.1 Weight of evidence for paternity 111
7.1.2 Prior probabilities 112
7.1.3 Calculating likelihood ratios 113
7.1.4 Multiple loci: the effect of linkage 117
7.1.5 s may be related to c but is not the father 119
7.1.6 Incest 120
7.1.7 Mother unavailable 121
7.1.8 Mutation 122
7.2 Other relatedness between two individuals 126
7.2.1 Only the two individuals profiled 126
7.2.2 Profiled individual close relative of target 127
7.2.3 Profiles of known relatives also available † 128
7.3 Software for relatedness analyses 129
7.4 Inference of ethnicity or phenotype † 131
7.5 Relatedness exercises 133

8 Other approaches to weight of evidence 135
8.1 Uniqueness 135
8.1.1 Analysis 136
8.1.2 Discussion 138

	8.2 Inclusion/exclusion probabilities	138
	8.2.1 Random man	138
	8.2.2 Inclusion probability of a typing system	139
	8.2.3 Case-specific inclusion probability	139
	8.3 Hypothesis testing †	141
	8.4 Other exercises	143
9	**Issues for the courtroom**	**145**
	9.1 Bayesian reasoning in court	145
	9.2 Some fallacies	146
	9.2.1 The prosecutor's fallacy	146
	9.2.2 The defendant's fallacy	147
	9.2.3 The uniqueness fallacy	148
	9.3 Some UK appeal cases	148
	9.3.1 Deen (1993)	148
	9.3.2 Dalby (1995)	149
	9.3.3 Adams (1996)	149
	9.3.4 Doheny/Adams (1996)	151
	9.3.5 Watters (2000)	153
	9.4 US National Research Council reports	154
	9.5 Prosecutor's fallacy exercises	155
10	**Solutions to exercises**	**157**
	Bibliography	**175**
	Index	**183**

Preface

Thanks are due to Kathryn Firth for drawing the figure on page 12 and to Karen Ayres and Renuka Sornarajah for providing helpful comments on a draft of the book. Lianne Mayor contributed some of the material for Section 7.2.3. Discussions with many colleagues and friends over more than 10 years have contributed to the ideas in this book: many forensic scientists have helped me towards some understanding of the laboratory techniques, and I thank Peter Donnelly for his stimulating comments on, suggestions about, and criticisms of my statistical ideas during the formative years of my interest in the field.

I am grateful to John Buckleton and Colin Aitken for sending me pre-publication manuscripts of Buckleton et al. (2004) and Aitken and Taroni (2004), respectively; at the time of writing, the published versions of these works had not appeared.

The statistical figures in this book were produced using R, a software package for statistical analysis and graphical display that has been developed by some of the world's leading statisticians. R is freely available for multiple platforms, with documentation, at www.r-project.org. Other diagrams have been created by the author using xfig, interactive drawing freeware that runs under the X Window System on most UNIX-compatible platforms, available at www.xfig.org.

1

Introduction

1.1 Weight-of-evidence theory

The introduction of DNA evidence at the end of the 1980s was rightly heralded as a breakthrough for criminal justice, but it had something of a "baptism of fire". In the media and in courts there was substantial controversy over the validity of the technology and the appropriate interpretation of the evidence.

DNA profiling technology has advanced impressively since then, and understanding by lawyers and forensic scientists of the appropriate methods for *interpreting* DNA evidence has also generally improved. Consequently, disputes about the accuracy and reliability of DNA evidence, and about its interpretation, have diminished in number and volume. However, the potential for crucial mistakes and misunderstandings remains. Although DNA evidence is typically very powerful, the circumstances under which it might not lead to satisfactory conclusions about identification or relatedness are not widely appreciated.

The primary goal of this book is to help equip a forensic scientist charged with presenting DNA evidence in court with guiding principles and technical knowledge for

- the preparation of statements that are fair, clear, and helpful to courts, and
- responding to questioning by judges and lawyers.

The prototype application is identification of the (single) culprit whose DNA profile was recovered from a crime scene, but we will also discuss profiles with multiple contributors, as well as paternity and other relatedness testing. The latter arise in both criminal and civil cases, as well as in the identification of human remains. We assume the setting of the US, UK, and Commonwealth legal systems in which decisions on guilt or innocence in criminal cases are made by lay juries, but the general principles should apply to any legal system.

Weight-of-evidence for Forensic DNA Profiles David Balding
© 2005 John Wiley & Sons, Ltd ISBN: 0-470-86764-7

We will introduce and develop a weight-of-evidence theory based on two key tenets:

1. The central question in a criminal trial is whether the defendant is guilty.

2. Evidence is of value inasmuch as it alters the probability that the defendant is guilty.

Although these tenets may seem self-evident, it is surprising how often they are violated. Focussing on the right questions clarifies much of the confusion that has surrounded DNA evidence in the past.

It follows from our tenets that evidential weight can be measured by likelihood ratios and combined to assess the totality of the evidence using the appropriate version of Bayes Theorem. We will discuss how to use this theory in evaluating evidence and give principles for, and examples of, calculating likelihood ratios, including taking into account relevant population genetic factors.

No theory ever describes the real world perfectly, and the analysis of forensic DNA profiles is a complex topic. It follows that the theory developed in this book cannot be applied in a naive, formulaic way to the practical situation faced by lawyers and forensic scientists in court. Nevertheless, a firm grounding in the principles of the theory provides

- grounds for deciding what information a clear-thinking juror needs in order to understand the strength of DNA profile evidence;

- the means to detect and thus avoid serious errors;

- a basis for assessing approximations and simplifications that might be useful in court;

- a framework for deciding how to proceed when the case has unusual features.

Fortunately, we will see that the mathematical aspects of the theory are not too hard. Of course, assessing some of the relevant probabilities – such as the probability that a sample handling error has occurred – can be difficult in practice, reflecting the real-world complexity of the problem. Further complications can arise for example in the case of mixed DNA samples (Section 6.5). However, the same simple rules and principles can give useful guidance in even the most complex settings.

Universal agreement is rare in the academic world, and there exist alternative theories of weight of evidence based on, for example, belief functions or fuzzy sets, rather than probabilities. The theory presented here is the most widely accepted, and its philosophical underpinnings are compelling (Bernardo and Smith 1994; Good 1991). It follows that whatever is actually said in court in connection with DNA evidence should not conflict with this theory.

There has been debate about the appropriateness in court of using numbers to measure weight of evidence. We only touch on this argument here (Section 6.3.3).

INTRODUCTION 3

It is currently almost a universal practice to accompany DNA evidence by some sort of numbers to try to measure its weight, and so we focus here on issues such as which numbers are most appropriate in court and how they should be presented.

1.2 About the book

Chapters 2, 3, and 9 are not scientifically technical and, for the most part, are not specific to DNA evidence. I therefore hope that lawyers dealing with scientific evidence, and forensic scientists not principally concerned with DNA evidence, will also find at least these chapters to be useful. Courtroom lawyers ignorant of the weight-of-evidence theory described in Chapters 2 and 3 should be as rare as theatre critics ignorant of Shakespeare, yet, in reality, I suspect that few are able to command its elegance, power, and practical utility.

I first set out the weight-of-evidence theory informally, via a simplified model problem (Chapter 2) and then more formally using likelihood ratios (Chapter 3). In Chapter 4, we briefly survey DNA-based typing technologies, starting with an introduction to autosomal[1] STR typing, emphasizing possibilities for typing error, then moving on to other DNA typing systems, and finishing with a brief digression to discuss fingerprint evidence. Next, we survey some population genetics theory relevant to DNA profile evidence (Chapter 5). These two chapters prepare us for calculating likelihood ratios for DNA evidence, which is covered in Chapters 6 (identification) and 7 (relatedness). In Chapter 8, we discuss some alternative probability-based approaches for assessing evidential strength: none of these methods is recommended but each has its merits, which should be understood and appreciated. In Chapter 9, I draw together ideas from the previous chapters and bring them to bear on the problem of conveying effectively, clearly, and fairly the weight of the DNA profile evidence to the court. To this end, we discuss some basic fallacies and briefly review the opinions of some UK and US legal and scientific authorities.

1.3 DNA profiling technology

For the most part, we will assume that the DNA evidence is summarized for reporting purposes as the lengths of short tandem repeat (STR) alleles at multiple (perhaps 10 or more) autosomal loci. The final result at four of the loci might be reported as

STR locus:	D18	D21	THO1	D8
Genotype:	14, 16	28, 31	9·3, 9·3	10, 13

in which each pair of numbers at a locus indicates the number of repeat units in the individual's two alleles. Although whole repeats are the norm, partial repeats

---

[1] The nuclear chromosomes excluding X and Y.

sometimes occur (Section 4.1.1); the profile represented here is homozygous for a THO1 allele that includes a partial repeat.

STRs now form the standard DNA typing technology in many countries. Currently in routine use in the United Kingdom and several other countries is an 11-locus system, including the sex-identifying locus Amelogenin, developed by the Forensic Science Service and known as *SGMplus*. CODIS is a 13-locus system developed by the FBI and widely used in the United States. The two systems are similar and indeed have eight loci in common (see Buckleton *et al.* 2004). PowerPlex® is a commercially available 16-locus STR typing system that contains the 13 CODIS loci, two pentanucleotide repeat loci, and Amelogenin.

The process of typing STR profiles is introduced in Section 4.1 but is not covered in great depth in this book. For further details emphasizing the CODIS system, see Butler (2001), and for a UK perspective, see Gill (2002). Rudin and Inman (2002) gives a general introduction both to technical and interpretation issues. Although we emphasize STR profiles, the principles emphasized below apply equally to any DNA profiling system. Interpreting profiles from the haploid parts of the human genome (the Y and mitochondrial chromosomes) raises special difficulties. These systems are introduced in Sections 4.2 and 4.3, and interpretation issues specific to them are discussed briefly in Section 6.4. In Section 4.5, we briefly discuss profiles based on single-nucleotide polymorphism (SNP) markers.

1.4 What you need to know already

Chapters 2, 3, and 9 have essentially no technical prerequisites. To follow Chapter 5, you should know already what an STR profile is, have a rudimentary genetics vocabulary (locus, allele, etc.), and know the basic ideas of Mendelian inheritance. In statistics, you should be familiar at least with the theory of the error in a sample estimate of a population proportion (binomial distribution). The reader with experience in calculating with probabilities will be at an advantage in Chapters 6 and 7, but few technical tools are required from probability theory. In Sections 5.4.1 and 8.3, familiarity with statistical hypothesis testing is assumed, but these sections are labelled with a †, which means that they can be skipped without adverse impact on your understanding of the remainder of the book. The sampling formula (5.16) will at first seem daunting to those without a mathematical background, but the simpler recursive form (5.6) can always be used to build up more complex formulas sequentially. I give examples of its use, which requires only an ability to add and to multiply, and with practice anyone should be able to use it without difficulty.

I do not provide a general introduction to statistics (for an introduction in forensic settings, see Aitken and Taroni 2004) and give only a brief introduction to population genetics (Section 5.1). I strongly believe that many complications and much confusion have arisen unnecessarily in connection with DNA evidence because of a failure to grasp the basic principles of assessing evidential weight. If one focusses on the questions directly relevant to the forensic use of DNA profiles, the number of ideas and techniques needed from statistics and population genetics will be small.

While the central ideas are not very difficult, inevitably there are special cases with their unique complexities. In addition, new ideas always take some time to absorb. Given some effort, this book should equip you with the basic principles for tackling any problem of interpreting forensic DNA evidence. The details of complex scenarios involving, for example, mixed profiles with missing alleles, will never be straightforward, and no book can replace the need for intelligence, care, and judgement on the part of the forensic scientist. The goal of this book is to complement these with some technical information and bring them to bear on the appropriate questions with guiding principles for assessing weight of evidence.

1.5 Other resources

Part of the reason for writing the book is to synthesize and extend in a coherent manner previous contributions to the forensic science and related literature by myself and co-authors. In particular, Chapter 3 is a development of Balding (2000), Section 7.1 extends the paternity section of Balding and Nichols (1995), and Section 8.1 is based on Balding (1999). Perhaps the most important feature of the book is the introduction of the population genetics sampling formula (Section 5.3) and its systematic application to various identification and relatedness problems. This draws in part on Balding and Donnelly (1995a), Balding and Nichols (1995), and Balding (2003) but some of the development is new here.

There are several other books that deal with the statistical interpretation of DNA and other evidence. Aitken (1995), soon to be superseded by Aitken and Taroni (2004), gives a thorough introduction to the statistical interpretation of scientific evidence in general, including DNA evidence among other evidence types. Robertson and Vignaux (1995) also deals with a range of evidence types and emphasizes interpretation issues from a lawyer's perspective, giving less attention to technical scientific aspects; for example, it does not discuss population genetics. Evett and Weir (1998) is perhaps closest to the present work, but the treatment of population genetics issues by these authors is very different from mine, as is their approach to introducing the relevant statistical issues. At the time of writing, Buckleton *et al.* (2004) is about to appear, offering a more extensive treatment of the interpretation issues raised by STR profile evidence.

As far as I can see, there is no major philosophical difference between myself and these authors: we all embrace the use of likelihood ratios and Bayes Theorem to evaluate evidence. We emphasize different aspects according to our individual perspectives, experience, and target audiences. The present book develops the weight-of-evidence theory in general and from an introductory level, and its approach to population genetics issues is unique, while remaining concisely focussed on DNA profile evidence, without extensive related material.

The December 2003 issue of *International Statistical Review* includes a series of papers dealing with the statistical interpretation of legal evidence, including a review of the interpretation of DNA evidence from a UK-based, historical perspective (Foreman *et al.* 2003). Other useful references, presented from a somewhat distinct viewpoint, are Kaye and Sensabaugh (2000, 2002). A widely used reference that has much useful background material but also, in my opinion, important flaws, is National Research Council (1996); my criticisms of it are outlined in Section 9.4. Charles Brenner's "Forensic Mathematics" website dna-view.com is a rich source of information and discussion, some of which are summarized in an encyclopedia article (Brenner 2003). Weir (2003) offers a more extensive, one-chapter summary of many issues.

2

Crime on an island

For most of this book, we will be thinking about one or other of the following problems:

- We have two matching DNA profiles, one from a known individual and one from a crime scene. How do we measure the weight of this evidence in favour of the proposition that the known individual is the source of the crime-scene profile?

- We have DNA profiles from two or more known individuals. Some of the relationships among them may be known, but at least one putative relationship is in doubt. For example, we may have a known mother and child, and a possible father-child relationship is in question. How do we assess the weight of the DNA evidence for or against the proposed relationship(s)?

We will see that there are many subtleties and potential complicating factors in answering these questions: even formulating the right questions precisely is not easy. We need to proceed in steps, starting with simplified problems and progressively adding more realistic features. In fact, before we even start to think about identification or relatedness evidence, we will look at some warm-up examples to get the brain into gear and into thinking about how to evaluate information using probabilities.

2.1 Warm-up examples

2.1.1 Disease testing: Positive Predictive Value (PPV)

Suppose that, although showing no symptoms, you decide to take a diagnostic test for a particular disease. Let us assume the following facts:

- about 1 in 1000 people with no symptoms will get the disease;
- the test is highly specific: the false-positive rate is 1% (= the proportion of positive outcomes among those who will not get the disease);
- the test is sensitive: the false-negative rate is 5% (= the proportion of negative outcomes among those who will get the disease);

The test result comes back positive. How worried should you be? What is the probability that you will get the disease?

The answer is that, even after the positive result from a reliable test, it remains unlikely that you will get the disease. The reason is illustrated in Figure 2.1. For every 100 000 individuals tested, on average 1094 will test positive, but of these only 95 will be true positives; that is, just under 9% of positive test results correspond to true positives and so there is still over 90% probability that you will not get the disease. The test is not valueless: it has increased your disease probability from 1 in 1000 to about 87 in 1000, a big relative increase but, because of the small starting value, the final value is still not large.

Of course, this simple analysis does not apply exactly to you or to me. In reality, there will be extra information, such as our age, sex, ethnicity, weight, and general state of health, that affect our chances of having any particular disease. However, the analysis reveals an insight that is broadly valid in real situations: we must consider both the accuracy of the test *and* the prevalence of true positives in interpreting the test results.

The disease-testing problem is analogous to the weight-of-evidence problem: there are two possible "states of nature", disease and no disease (compare with guilty and not guilty), and there is a diagnostic test that is reliable but can occasionally fail

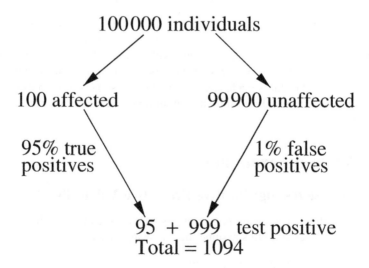

Figure 2.1 Schematic solution to the disease-testing problem. The proportion of true positives among those who test positive is 95/1094, or about 8·7%.

CRIME ON AN ISLAND

(compare with a DNA profile test, which can occasionally result in a "chance" match or a laboratory or handling error). Our analysis of disease testing leads to results that at first are counter-intuitive for many people. Nevertheless, the logic of the analysis is compelling, and its implications are now universally accepted for, for example, public health policy.

We will see that the analogous reasoning for DNA profile evidence also leads to some surprising conclusions and, after a struggle during the 1990s, these are also becoming accepted in the courts. In public health, the probability of having the disease given a positive test result is called the "positive predictive value" (PPV) of the test. It is a difficult quantity because it depends on the disease prevalence – and this will vary, for example, according to ethnic group and occupation. In contrast, the false-positive and false-negative rates are easier to work with because they can be measured in the laboratory. However, it is the PPV that is relevant for many questions, not the error rates alone. Similarly, in forensic identification, it is not the "match probability" for the DNA profile test that ultimately matters, but the equivalent of the PPV, and this is a difficult quantity because it depends, for example, on the other evidence in the case.

2.1.2 Coloured taxis

Next, we consider an example of the sort of reasoning that led to the PPV for disease testing, but in the setting of a simplified evidence assessment problem. Suppose that 90% of the taxis in the town are green and the rest are blue. According to an eyewitness, the perpetrator of a "hit-and-run" traffic offence was driving a blue taxi. Assume that the eyewitness testifies honestly but may have made a mistake about the colour of the taxi: it was dark at the time, and tests indicate that, under these conditions, eyewitnesses mistake blue taxis for green, and vice versa, about 1 time in 10.

What is the probability that the taxi was really blue?

The false-positive and false-negative rates are both 10%, but we have seen that these do not immediately provide the answer we seek. As a "thought experiment", consider 100 such eyewitness reports. About 18 taxis will be reported to be blue, but only 9 of these are truly blue (Figure 2.2). So, after hearing an eyewitness report that the taxi was blue, the probability that it really was blue is 50%: the "diagnostic" information that the taxi was blue exactly balances the background information that blue taxis are rare.

Had the eyewitness reported that the taxi was green, then the diagnostic information would support the background information to give an overall 81/82, or $\approx$ 99%, probability that the taxi really was green. This is a well-known example for evidence scholars and causes difficulty because it seems to imply that green-taxi drivers who offend will always get caught, yet blue-taxi drivers can drive dangerously with impunity. We do not discuss this conundrum further here but note that it is not unreasonable to require more evidence to be persuaded that a rare event has occurred than for a common event.

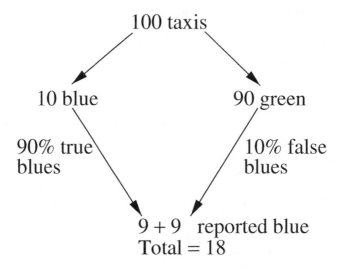

Figure 2.2 Schematic solution to the taxi problem. The probability that a taxi reported to be blue really was blue is 9/18, or 50%.

2.2 Rare trait identification evidence

The world is a complicated place. One of the methods scientists use to try to understand it is to investigate simple models. Seeing what happens when simple models are tweaked can suggest insights into how the real world works. Eventually, bells and whistles can be added to the models to make them a little closer to reality, but it is often the simplest models that give the most profound insights.

We will take this approach for assessing the weight of DNA evidence. An actual case involves many complications:

- How did the suspect come to be arrested?
- How many previous suspects were investigated and excluded?
- What were the possibilities for a contamination error?
- Are any of the suspect's close relatives possible culprits?

and many more. We cannot cope with all these complications at once. Instead, we will follow Eggleston (1983) and start with an imaginary crime on an imaginary island where life and crime are much simpler than in our world. Although unrealistic, analysis of the "island problem" leads to profound insights. We will gradually add more features to bring the island closer to the real world, and in doing so, we will learn new lessons about evidential weight.

2.2.1 The "island" problem

Consider a rare trait: let's call it Υ. It could be a DNA profile – there is no need to be specific at this stage – but whether someone has Υ is not immediately apparent;

CRIME ON AN ISLAND

it requires a test that, for the moment, we will assume is perfectly accurate. A crime is committed on a remote island with a population of, say, 101. At first, there are no clues, and everyone on the island is equally under suspicion. Then, it is learned that the culprit must possess Υ, and a suspect is identified who is tested and found to have Υ. How convinced should we be that the suspect is the culprit?

The answer to this question depends on, among other factors, how rare Υ is. Suppose that the suspect and the culprit (who may or may not be the same person) are the only people on the island whose Υ-status is known. A recent survey on the nearest continent, however, indicated that 1% of the population has Υ and we assume that this rate holds, on average, on the neighbouring islands. We also assume that the probability that another islander has Υ is unaffected by the knowledge that the suspect has it.

The facts of the island problem are summarized on page 12. A number of false "solutions" have been presented in the evidence literature (for a discussion, see Balding and Donnelly 1995a; Eggleston 1983). The correct solution is sketched in Figure 2.3.

Even though Υ is rare, there is still only a 50% chance that the suspect is the culprit. In addition to the culprit, we expect one innocent person on the island to have Υ, so that there is only 1 chance in 2 that we have identified the right person. Just as for the case of the taxi reported to be blue, the background information that any particular individual is unlikely to be the true culprit is exactly balanced here by the diagnostic information of the Υ-match.

2.2.2 A first lesson from the island problem

It is the probability that the defendant is the culprit given a match of Υ-states – the analogue of the PPV for disease testing – that is directly relevant to the juror's decision. Like the PPV, this probability is, in reality, a slippery quantity to work with. Much of the remainder of this book is devoted to grappling with it. But we must walk before we can run, and for the moment, we will continue to work in the simplified island setting in which we can calculate the probability of guilt. Here, we are given that there is no other evidence and that all islanders are initially equally under suspicion, so we are permitted to ignore factors that complicate real-world crime investigations, like the state of health and distance from the crime scene of a possible suspect.

In general, if there are N people on the island other than the suspect, and the probability that anyone of them has Υ is p, then using a special case of a more general "weight-of-evidence" formula, given in Section 3.2 below, we have

$$P(G \mid E) = \frac{1}{1 + N \times p}. \tag{2.1}$$

The notation "|" is mathematical shorthand for "given", and we use P for "probability", G for "guilty" and E for "evidence". Thus, $P(G \mid E)$ is a concise shorthand

The island problem: facts summary

- All 101 islanders are initially equally under suspicion.
- The culprit has ϒ.
- The suspect has ϒ.
- The ϒ-states of the other islanders are unknown.
- We expect on average about 1 person in 100 to have ϒ.

What is the probability that the suspect is the culprit?

for "the probability of guilt given the evidence". Mathematical notation is remarkably ink-efficient; that means it can seem opaque to those unfamiliar with it, but remember that it is just efficient shorthand, and it always stands for something that can be expressed (at much greater length) in words.

If $N = 100$ and $p = 1/100$, then

$$P(G \mid E) = \frac{1}{1 + 100 \times 1/100} = \frac{1}{2}.$$

The 1 in the denominator corresponds to the suspect, and the $100 \times 1/100$ corresponds to the (expected) one other ϒ-bearer on the island.

CRIME ON AN ISLAND

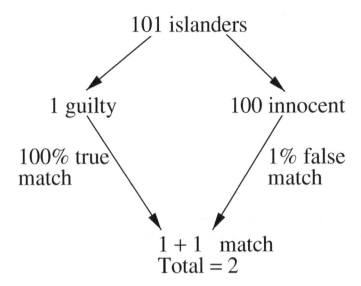

Figure 2.3 Schematic solution to the island problem. The probability that a matching individual is the true culprit is 1/2, or 50%.

Although it follows from equation (2.1) that the rarer the Υ is, the higher is the probability that the suspect is the culprit, the strength of the overall case against the suspect depends on *both* the rarity of Υ (i.e. on p) and on the number of alternative possible culprits N – just as the probability of getting the disease following a positive test result depends on *both* the error rates and the prevalence of the disease.

Lesson 1 *The fact that Υ is rare (i.e. p is small) does not, taken alone, imply that the defendant is likely to be guilty.*

The unrealistic aspects of the island problem are immediately apparent. Nevertheless, Lesson 1 is "robust": when we change the problem to make it more realistic, Lesson 1 still applies (see Section 3.4.1).

2.3 Making the island problem more realistic

Let us change some aspects of the island problem to investigate how different factors affect evidential weight. To avoid getting too bogged down in complications, we will investigate different factors one at a time.

2.3.1 Uncertainty about p

In practice, we will not know p exactly; there will be uncertainty about its value, which we can express as a variance, σ^2. The island problem formula (2.1) applies when $\sigma^2 = 0$, corresponding to perfect knowledge about p. The more general formula (see Section 3.5.2 for a derivation) is

$$P(G \mid E) = \frac{1}{1 + N(p + \sigma^2/p)}, \qquad (2.2)$$

which is always less than (2.1). We therefore have immediately the following important conclusion, which, like Lesson 1, turns out to be robust.

Lesson 2 *Uncertainty about the value of p does not "cancel out". Ignoring this uncertainty is always unfavourable to defendants.*

Numerical illustration: uncertainty about p

In the original island problem of Section 2.2.1, $N = 100$ and $p = 0.01$, and the probability of guilt given the evidence is

$$P(G \mid E) = \frac{1}{1 + N \times p} = \frac{1}{1 + 100 \times 0.01} = 50\%.$$

If, now, we suppose that there is some uncertainty about p, say $p = 0.01 \pm 0.005$ (i.e. $\sigma = 0.005$), then the probability of guilt is

$$\begin{aligned} P(G \mid E) &= \frac{1}{1 + N \times (p + \sigma^2/p)} \\ &= \frac{1}{1 + 100 \times (0.01 + (0.005)^2/0.01)} \\ &= \frac{1}{1 + 1 + 0.25} \\ &= \frac{4}{9} \approx 44\%. \end{aligned}$$

So, failing to acknowledge the uncertainty about p overstates the probability of guilt; here, 50% instead of the correct 44%.

CRIME ON AN ISLAND 15

Uncertainty about p can arise for a number of reasons. If knowledge about p comes from a survey, then there is uncertainty due to sampling: a different sample would have led to somewhat different results. Moreover, there is always uncertainty due to the possibility that the sample is unrepresentative (e.g. the islanders may not be similar, in important respects, to the population of the continent).

We will see in Chapter 6 that uncertainty about p is crucial to the correct interpretation of DNA evidence: uncertainty enters not only because of sampling error but also because DNA profile proportions vary among ethnic/religious/social groups and we never know exactly which is the correct reference group in a particular case, or what are the profile proportions in the relevant groups. Although we can recognize human groupings, and give them labels such as "Arab", "Hispanic", "Scandinavian", and so forth, because of migrations, intermarriages, and imprecise boundaries, we cannot define them precisely. The rarity of a DNA profile is assessed in part on the basis of data from "convenience" samples rather than scientific random samples, and in part on the basis of population genetics theory, which holds at best only approximately in actual human populations. In practice, then, the σ^2/p in equation (2.2) is often much larger than the p, so that ignoring the former can be much more important than in the illustration above.

2.3.2 Uncertainty about N

The size N of the island population, excluding the suspect, may also be unknown and uncertainty about its value also affects the probability that the suspect is guilty. In this case, however, the news is better: the effect of uncertainty about N is usually small and ignoring it tends to favour the suspect. For example, if the island population size is expected to be N, but it could be either $N-1$ or $N+1$, each with probability ϵ, then the island problem formula (2.1) becomes

$$P(G \mid E) \approx \frac{1}{1 + Np(1 - 4\epsilon/N^3)} > \frac{1}{1 + Np}. \qquad (2.3)$$

See Section 3.5.2 for a further details.

> **Lesson 3** *Uncertainty about N, the number of alternative possible suspects, also does not "cancel out", but the effect of ignoring this uncertainty is usually small and tends to favour defendants.*

2.3.3 Possible typing errors

Now, forget uncertainty, and assume again that we know both N and p exactly. Suppose, however, that there is a probability ϵ_1 that an individual who does not have Υ will be wrongly recorded as having it (i.e. a false-positive), and probability ϵ_2 for the other, false-negative, error. We assume that these probabilities apply for typing both the suspect and the crime-scene profiles and that errors occur independently.

These assumptions are still unrealistically simple, but they allow a first insight into how the possibility of error affects evidential weight. The exact formula is a little complex, and we defer its derivation until Section 3.5.4 below. But an approximation appropriate here is given by

$$P(G \mid E) = \frac{1}{1 + N(p + \epsilon_1)^2/p}. \tag{2.4}$$

Notice that, to a first approximation, ϵ_2 is irrelevant: since we have observed a match, the probability of an erroneous non-match isn't important.

Formula (2.4) suffices for another important, robust lesson.

Lesson 4 *The overall weight of evidence involves adding together the probability of a "chance match" and the probability of a match due to a typing error.*

Numerical illustration: typing errors

If we suppose that there is a probability $\epsilon_1 = 0{\cdot}005$ that any islander without Υ will be wrongly recorded as a Υ-bearer, then the probability of guilt is

$$P(G \mid E) \approx \frac{1}{1 + N \times (p + \epsilon_1)^2/p}$$

$$= \frac{1}{1 + 100 \times (0{\cdot}01 + 0{\cdot}005)^2/0{\cdot}01}$$

$$= \frac{1}{1 + (1{\cdot}5)^2}$$

$$= \frac{1}{3{\cdot}25} \approx 31\%.$$

So, ignoring the possibility of typing error leads to overstatement of the probability of guilt; here, 50% instead of the correct 31%.

Laboratory and handling errors for DNA profile evidence are discussed further in Section 3.4.4. In general, the value of ϵ_1 is difficult to assess, usually more difficult than p. This is because error rates are small, and so very large trials are required to assess them. Worse, any series of trials will never be able to mimic the

CRIME ON AN ISLAND 17

exact circumstances of a particular case, including the amount of DNA recovered and its level of contamination and/or degradation. Ultimately, it is for the jury in criminal trials to assess the probability that some error has occurred, on the basis of the evidence presented to it. It is important that courts be given some idea of what errors are possible, how likely they are in the present case, and what effect possible errors have on evidential weight.

2.3.4 Searches

Forget, for the moment, uncertainty and errors and focus on a new issue. The question of how the suspect came to the attention of the island crime investigators has so far been ignored. This is not as unrealistic as it may at first appear: in practice, suspects are often identified on the basis of a combination of factors such as previous convictions, suspicious behaviour, criminal associates, and so forth. Such reasons may not form part of the evidence presented in court, in which case, as far as a juror is concerned, the defendant "just happened" to come to the attention of the authorities. Further, the legal maxim "innocent until proven guilty" is usually interpreted to mean that, before the evidence is presented in court, the defendant should be regarded as being just as (un)likely to be guilty as anybody else in the population.

Suppose now that the suspect was identified on the basis of a search for Υ-bearers: the islanders were examined in random order until the first one with Υ was found. This individual was then accused of the crime. Besides the facts listed in the summary on page 12, we now have the additional information that, say, k islanders have been investigated and found not to have Υ.

In the original island problem, the reasons for first identifying the suspect are not given. Is $P(G \mid E)$, the probability that the suspect is the culprit, higher or lower given the extra information that s/he was identified following a search? There seem to be two principal arguments for believing that it should be lower:

(i) the fact that the suspect was initially just one person in a random sequence of individuals searched means that s/he is *a priori* less likely to be the culprit;

(ii) if you set out to find a suspect who has Υ, then the fact that the suspect has Υ is unsurprising and therefore of little or no evidential value.

It turns out that the probability that the suspect is guilty is *higher* following a search than in the original island problem. In fact,

$$P(G \mid E) = \frac{1}{1 + (N-k)p}, \qquad (2.5)$$

which is greater than (2.1). Many people, particularly scientists, find this result counter-intuitive, perhaps because of the two arguments given above.

Argument (i) is easily dismissed: "innocent until proven guilty" implies that every defendant should be treated, before the evidence is presented, as just another member of the population. This view is incorporated into the island problem by initially regarding every islander as equally under suspicion. So, there should be nothing different about a suspect identified on the basis of a search.

To see that argument (ii) is wrong, think about the case that the suspect was the last person to be searched: everyone else was inspected and found not to have Υ. Then, the fact that the suspect has Υ is overwhelmingly strong evidence against him/her, as is reflected by the value $P(G \mid E) = 1$ obtained in equation (2.5) when $N = k$. The key is to keep attention fixed on the relevant question, which is not

"how likely is it that I will find a Υ-bearer if I look for one?",

but

"given that a Υ-bearer has been found, how strong is the evidence against this individual?".

The reason behind the correct formula is that each individual found not to have Υ is excluded from suspicion (remember that we are ruling out test error here). The removal of this individual leaves a smaller pool of possible culprits, and hence, each remaining person in the pool becomes (slightly) more likely to be guilty. Notice that if the first person is found to have Υ, so that $k = 0$, then the original island problem formula is recovered. The fact that a search was intended, or whatever else was in the mind of the crime investigator, makes no difference to the strength of the evidence.

Lesson 5 *In the case of a search of possible culprits to find a "match" with crime-scene evidence, the longer the search (i.e. the more individuals found not to match) the stronger the evidence against the one who is finally found to match.*

The identification of suspects through a search of a DNA profile database is closely related to the island problem search and is discussed below in Section 3.4.5.

2.3.5 Other evidence

In the island problem, we assumed in effect that there was no evidence other than the Υ-evidence. In practice, of course, even if there is no further evidence that is directly incriminating, there will be background information presented to the jury, such as the location, time, and nature of the crime, that makes some individuals more plausible culprits than others.

We write w_i for the weight of the non-Υ evidence against person i, *relative to its weight against the suspect*. In the original island problem, each w_i was equal to one. A value of w_i greater than one would indicate that, ignoring the Υ-evidence, individual i is more likely to be the culprit than is the suspect. As an example, suppose that, other than the information about Υ, the evidence consists of the

CRIME ON AN ISLAND 19

location of the crime and the locations of the homes of all the islanders. A juror may reason that individuals who live near to the crime scene are *a priori* more likely to be the culprit than, say, individuals who live on the other side of the island. Such an assessment can be reflected by values of w_i greater than one for those who live nearer to the crime scene than the suspect, and less than one for those who live more distant than the suspect.

When other evidence is taken into account, the island problem formula (2.1) becomes

$$P(G \mid E) = \frac{1}{1 + p \times \sum_{i=1}^{N} w_i}, \qquad (2.6)$$

in which we introduce the mathematical symbol $\sum$ to denote summation, so that, for example,

$$\sum_{i=1}^{3} w_i \equiv w_1 + w_2 + w_3.$$

If $w_i = 1$ for all i, then $\sum_{i=1}^{N} w_i = N$, and the formula reduces to the original island problem formula.

The role of the w_i in connection with assessing DNA evidence is discussed further in Section 3.4.2.

2.3.6 Relatives and population subdivision

Even though we are initially ignorant about who on the island has Υ, the observation that the suspect has it can be informative about whether other individuals, such as relatives and associates, also have Υ. For example, if the population of the island is divided into "easties" and "westies", then the fact that the suspect, an eastie, has Υ may make it more likely that other easties also have Υ.

In the island problem, we assumed that Υ possession for different individuals was independent, so that the Υ-status of the suspect carries no information about the Υ-states of other individuals. In practice, however, this "learning" effect can be important, particularly for DNA profile evidence. Any particular DNA profile is very rare, but once that profile is observed, it becomes more likely that other people, especially among the individual's relatives or ethnic group, also have it. What matters in practice for DNA profile evidence is not p, the overall frequency of the profile, but the probabilities of the other possible culprits having the profile *given that* the suspect has it.

We will write r_i for the *match probability* for possible culprit i, which is the probability that i has Υ, *given that* the suspect has Υ. Some writers on DNA evidence use "match probability" to denote the relative frequency of Υ in some population. This is incorrect because the concept of "match" involves two individuals, not one. In the original island problem, we assumed independence of Υ-states so that r_i does equal p in that setting. For DNA profile evidence in real populations, however, relatedness and population subdivision mean that the match probability r_i exceeds the profile relative frequency p, often substantially. Thus, confusing r_i with p is detrimental to defendants (see Chapter 6).

When these effects are taken into account, the island problem formula (2.1) becomes

$$P(G \mid E) = \frac{1}{1 + \sum_{i=1}^{N} r_i}. \tag{2.7}$$

If all the r_i are equal to p, then $\sum_{i=1}^{N} r_i = N \times p$, and (2.1) is recovered.

> **Lesson 6** *The strength of identification evidence based on an inherited trait depends (often strongly) on the degree of relatedness of the defendant with the other possible suspects.*

2.4 Weight-of-evidence exercises

Solutions start on page 157.

1. Suppose that weather forecasts are accurate: the statistics show that rain had been forecast on about 85% of days that turned out to be rainy; whereas about 90% of dry days had been correctly forecast as dry. (To make matters simple, each day's weather has one of two states, "rainy" or "dry"). It follows that if rain is forecast, you may as well cancel the picnic, it is very likely to rain. Right or wrong?

    Assuming the above accuracy figures, what is

    (a) the probability of rain following a forecast of rain for Manchester, where rain falls, let's suppose, 1 day in 5?

    (b) the probability of rain following a forecast of rain for Alice Springs, where rain falls, say, 1 day in 100?

2. A diagnostic test for a latent, infectious medical condition has a false-positive rate of 2 in 1000 and a false-negative rate of 8 in 1000. Suppose that the prevalence of the condition in the general population is 1 in 10 000.

    (a) What is the probability that someone who tests positive has the condition?

    (b) Repeat (a), now assuming that the person tested comes from a high-risk group for which the prevalence is 1 in 250.

CRIME ON AN ISLAND

3. Suppose that the island has a population of 1001, all initially equally under suspicion. The suspect s is tested and found to have Υ, whereas the Υ-states of the other 1000 inhabitants are unknown. Given that an islander has Υ, the probability that a particular unrelated islander i also has it is $r_i = 5 \times 10^{-6}$, while $r_i = 2 \times 10^{-4}$ when i is a cousin of s and $r_i = 0.015$ for a sibling. Ignore here the possibility of testing error and uncertainty about the r_i.

   (a) Suppose that s is known to have no relatives on the island. What is the probability that he is the culprit?

   (b) Repeat (a), but now it is known that s has only one sibling, a brother named b. In addition, s has 20 cousins, but no other relatives, on the island.

   (c) Repeat (b), but now b agrees to be tested for Υ and is found not to have it.

   (d) As a juror in the case, what assessment would you make about evidential strength if you were not given any information about the numbers and locations of the relatives of s?

3

Assessing evidence via likelihood ratios

There are many factors beyond those discussed in Chapter 2 that could be introduced onto the island in order to investigate their effects on evidential strength:

- What if the culprit is not, after all, the source of the DNA obtained from the crime scene?

- What about the fact that the suspect failed to produce a convincing alibi?

- What if the police always accuse this suspect of every crime that occurs on the island?

- What if only a few islanders could have visited the crime scene during the time of the offence?

We will not pursue these now. Instead, we will now leave the island and introduce a general formula for quantitatively assessing evidence. Although we have given some intuitive explanation of formulas (2.1) through (2.7), we have not yet explained how to obtain such formulas. We give some explanation here; for a more advanced analysis, incorporating some of the factors listed above, see, for example, Balding and Donnelly (1995a) and Dawid and Mortera (1996).

3.1 Likelihood ratios

The match probability r_i, defined on page 19, is a special case of a quantity called a *likelihood ratio*. We first introduce likelihood ratios in general and later turn to likelihood ratios for DNA and other identification evidence.

ASSESSING EVIDENCE VIA LIKELIHOOD RATIOS

Consider data D that are potentially informative about two rival (i.e. mutually exclusive) hypotheses, H and G. The *likelihood ratio* $R_{H,G}(D)$ is the ratio of the probabilities of the data under the hypotheses:

$$R_{H,G}(D) = \frac{P(D \mid H)}{P(D \mid G)}.$$

It does not matter whether it is H on top (numerator) and G below (denominator), or vice versa, as long as it is clear which has been used. A likelihood ratio can be defined for any kind of data (or *evidence* or *information*). For example, Buckleton et al. (2004) note that absence of evidence (e.g. that a search of the defendant's home failed to find anything incriminating) is itself evidence for which a likelihood ratio can, in principle, be calculated.

In the disease-testing example of Section 2.1.1, the data is the fact that the test result was positive. This has probability 0·95 if the hypothesis "affected" (A) is true and probability 0·01 if "unaffected" (U) holds. The likelihood ratio is thus

$$R_{A,U}(+\text{ve test}) = \frac{P(+\text{ve test} \mid A)}{P(+\text{ve test} \mid U)} = 95.$$

The positive test result supports hypothesis A much more than U. However, recall that because A may be *a priori* very unlikely, it does not necessarily follow that D suffices to make A more likely than U.

The value of $R_{H,G}(D)$ is a measure of the weight of evidence conveyed by D for H *relative to* its weight for G:

> $R_{H,G}(D) > 1$ means that D is more likely if H is true than if G is true. Whatever weights we previously assigned to the two hypotheses, observing D should lead us to increase our belief that H is true relative to our belief in G.

If H and G are not exhaustive, it may be that D is very unlikely under both, but highly likely under a third hypothesis. In that case, the probabilities for H and G may both decline as a result of observing D, but that for G will decline relatively further. However, if H and G are exhaustive (one of them must be true), then $R_{H,G}(D) > 1$ does imply that observing D increases the probability of H and decreases that for G.

If there are several items of data, we may wish to consider the likelihood ratio for data D given that data D' has already been taken into account. In this case, we can define

$$R_{H,G}(D \mid D') = \frac{P(D \mid H, D')}{P(D \mid G, D')}.$$

In the disease-testing example, D' may include background information about the person being tested, for example, their sex, age, and weight. This information may affect the likelihood ratio, and it usually affects the prior probabilities of the two hypotheses.

In Section 6.1, there is a further discussion of likelihood ratios as measures of evidential weight.

3.2 The weight-of-evidence formula

> **Definition** Let E_d denote the evidence to be assessed, and write E_o for the background information (including evidence that has already been assessed). Write C for the true culprit, and i for a possible suspect (other than the defendant, s). We write $R_{i,s}(E_d \mid E_o)$ for the **likelihood ratio**:
> $$R_{i,s}(E_d \mid E_o) = \frac{P(E_d \mid C = i, E_o)}{P(E_d \mid C = s, E_o)}. \tag{3.1}$$

Most authors define the likelihood ratio the other way around (i.e. with the top and bottom lines interchanged). Either definition is acceptable given the obvious adjustments to formulas. The conventional definition has the advantage that the likelihood ratio corresponding to strong evidence can often be expressed as a whole integer, whereas for us $R_{i,s}(E_d \mid E_o)$ is a small fraction when the new evidence E_d provides strong evidence against $C = i$, beyond that already incorporated in E_o. However, our definition simplifies many formulas and so is useful for the purposes of this book. In particular, it has the advantage that $R_{i,s}(E_d \mid E_o)$ can often be interpreted as a conditional probability, the *match probability* (see Chapter 6).

> **Definition** The **other evidence ratio** $w_{i,s}(E_o)$, introduced informally on page 19, is the probability that $C = i$, divided by the probability that $C = s$, both evaluated in the light of E_o but ignoring E_d. That is,
> $$w_{i,s}(E_o) = \frac{P(C = i \mid E_o)}{P(C = s \mid E_o)}. \tag{3.2}$$

We will often write R_i and w_i instead of $R_{i,s}(E_d \mid E_o)$ and $w_{i,s}(E_o)$ if it is clear what evidence E_d is being evaluated, what background information E_o is being assumed, and that s is the defendant against whom i is to be compared.

Putting together the factors discussed in Sections 2.3.5 and 2.3.6, with the more general R_i replacing the r_i, we obtain the following:

The weight-of-evidence formula

$$P(G \mid E_d, E_o) = \frac{1}{1 + \sum w_i R_i}, \quad (3.3)$$

where the summation is over all i in $\mathcal{P}$, the population of alternative possible culprits.

Formula (3.3) is a special case of a result in probability theory known as *Bayes Theorem*, in honour of the eighteenth century clergyman Thomas Bayes. See an introductory probability textbook for more general formulations. I like to call this special case the "weight-of-evidence formula", but this appellation is not standard.

3.2.1 Application to the island problem

In the island problem setting, the evidence E_d can be summarized by

E_d = "both suspect (defendant) and culprit are observed to have Υ".

Since typing error is assumed impossible, E_d implies that suspect and culprit do both have Υ, in which case each likelihood ratio R_i is equivalent to the match probability r_i. The principle distinction between the two is that the likelihood ratio allows the possibility for the evidence to have arisen by other means, such as handling error or fraud.

If the suspect is not the culprit, then we have observed two Υ-bearers on the island: the culprit and the suspect. Under the assumptions introduced on page 11, the probability that any two individuals both have Υ is $p \times p = p^2$. On the other hand, if the suspect is the culprit, then we have observed only one Υ-bearer and this observation has probability p. The likelihood ratio for any possible culprit i is thus

$$R_i = \frac{p^2}{p} = p. \quad (3.4)$$

Substituting this value into equation (3.3), we recover (2.6), and in the case that all the w_i are equal to one, we once again obtain the original island problem formula (2.1).

3.2.2 The population $\mathcal{P}$

The population $\mathcal{P}$ is assumed to include all the possible sources of the crime stain other than s. Although $\mathcal{P}$ should include all realistic alternative suspects, there is some flexibility as to how many extra individuals are included. Often it might be appropriate to include in $\mathcal{P}$ all men aged, say, between 16 and 65 living within, say, 1 hour driving time of the crime scene. Alternatively, $\mathcal{P}$ might include all adult male residents of the nation in which the crime occurred. However, $\mathcal{P}$ could include everyone on earth except s, if desired: for a crime committed in Marrakesh, the value of w_i will be very close to zero when i is a resident of Pyongyang. This individual can be included in $\mathcal{P}$, but the error resulting from simply omitting from $\mathcal{P}$ all the residents of Pyongyang, and indeed all individuals who reside far from Marrakesh, will usually be negligible.

Since $\mathcal{P}$ can be large, it may seem impractical to compute a separate R_i for *every* member of $\mathcal{P}$. In practice, however, there will be groups of individuals relative to whom the evidence E_d bears the same weight against s, and thus for whom R_i will take the same value. Partitioning $\mathcal{P}$ into these groups can simplify (3.3).

For DNA evidence, there will typically be a small number of groups within which individuals bear approximately the same degree of relatedness to s. For example, the population $\mathcal{P}$ of alternative culprits may be partitioned into

- identical (monozygote) twins of s,
- siblings (including dizygote twins),
- parents and offspring of s,
- 2nd degree relatives such as uncle, nephew, and half-siblings,
- 3rd degree relatives (cousins),
- unrelated – same population, same subpopulation,
- unrelated – same population, different subpopulation, and
- unrelated – different population.

To avoid overstating the evidence against s, if the value of R_i varies within a group, then the largest value should be applied for all members of the group. Then the denominator of (3.3) can be rewritten with just one term for each of these groups (Section 3.5.3). This gives a simpler formula that provides only a lower bound on the probability of guilt, but the bound will often be adequate for practical use.

The notions of "population" and "subpopulation", although intuitive and widely used, are difficult to define precisely in practice and hence their usefulness may be disputed. These, and possibly other groups in the above list, can be combined, leading to an even simpler formula that, if we continue to apply the largest R_i to all members of the combined group, gives a bound that is cruder but may still be satisfactory for use in court.

3.3 General application of the formula

3.3.1 Several items of evidence

When the evidence to be assessed, E_d, consists of two items, say E_1 and E_2, the likelihood ratio can be calculated in two equivalent ways, corresponding to the two possible orderings of E_1 and E_2:

$$R_i(E_1, E_2 \mid E_o) = R_i(E_2 \mid E_o)R_i(E_1 \mid E_2, E_o)$$
$$= R_i(E_1 \mid E_o)R_i(E_2 \mid E_1, E_o), \quad (3.5)$$

where, for example,

$$R_i(E_2 \mid E_1, E_o) = \frac{P(E_2 \mid C = i, E_1, E_o)}{P(E_2 \mid C = s, E_1, E_o)}, \quad (3.6)$$

is the likelihood ratio for evidence E_2 when E_1 has already been assessed and is now included with the background information, E_o. Further,

$$w_i(E_o)R_i(E_1 \mid E_o) = \frac{P(C = i \mid E_o)}{P(C = s \mid E_o)} \frac{P(E_1 \mid C = i, E_o)}{P(E_1 \mid C = s, E_o)}$$
$$= \frac{P(C = i \mid E_1, E_o)}{P(C = s \mid E_1, E_o)}$$
$$= w_i(E_1, E_o).$$

Thus,

$$w_i(E_o)R_i(E_1, E_2 \mid E_o) = w_i(E_1, E_o)R_i(E_2 \mid E_1, E_o),$$

so that applying the weight-of-evidence formula to E_1 and E_2 together, given background information E_o, gives the same result as applying it to E_2 when E_1 is included with E_o. The result is also the same when the formula is applied to E_1 when E_2 is included with E_o.

If you are slightly confused at this point, do not panic! There are two simple take-home lessons that follow from the above manipulations:

Take-home lesson (i): dependent items of evidence

Apparently, strong evidence may be of little value if it merely replicates previous evidence. If E_1 and E_2 are highly correlated (e.g. matches at tightly linked genetic loci, or statements from two friends who witnessed the crime together and discussed it afterwards), then $R_i(E_1 \mid E_2, E_o)$ and $R_i(E_2 \mid E_1, E_o)$ may both be close to one (i.e. little evidential weight) even though both $R_i(E_1 \mid E_o)$ and $R_i(E_2 \mid E_o)$ are small (strong evidence). In this case, (3.5) implies that the joint weight of the two pieces of evidence is about the same as the weight of either piece of evidence taken alone. On the other hand, if the items of evidence are independent, then

$$R_i(E_1, E_2 \mid E_o) = R_i(E_1 \mid E_o) R_i(E_2 \mid E_o). \tag{3.7}$$

For example, suppose that two eyewitnesses give essentially the same testimony (E_1 and E_2), and that for either one of them a juror assesses that the evidence is 10 times more likely if s is the culprit than if i is guilty, so that

$$R_i(E_1 \mid E_o) = R_i(E_2 \mid E_o) = 0 \cdot 1.$$

In one case a juror may assess that the witnesses have discussed their evidence beforehand, and smoothed over any differences, so that the second witness adds nothing to the first. Then $R_i(E_2 \mid E_1, E_o) = 1$ and

$$R_i(E_1, E_2 \mid E_o) = R_i(E_2 \mid E_1, E_o) R_i(E_1 \mid E_o) = 1 \times 0 \cdot 1 = 0 \cdot 1.$$

In another setting, the two witnesses may be regarded as independent:

$$R_i(E_1, E_2 \mid E_o) = R_i(E_2 \mid E_1, E_o) R_i(E_1 \mid E_o) = 0 \cdot 1 \times 0 \cdot 1 = 0 \cdot 01.$$

Thus, the likelihood ratio captures the intuition that two independent pieces of evidence are jointly more powerful (here $R = 0 \cdot 01$) than two pieces of evidence that are essentially just replicates of the same information (here $R = 0 \cdot 1$).

Take-home lesson (ii): order of evidence doesn't matter

The order in which different pieces of evidence are assessed does not matter, provided that the likelihood ratio is calculated correctly, nor does the way that evidence is grouped together for assessment purposes. The likelihood ratio for multiple items of evidence can be built up by sequentially considering each item, in any order, but at each step all previously considered items must be included with the background information. For example,

$$R_i(E_1, E_2, E_3 \mid E_o) = R_i(E_3 \mid E_o) R_i(E_2 \mid E_3, E_o) R_i(E_1 \mid E_2, E_3, E_o)$$
$$= R_i(E_1 \mid E_o) R_i(E_2 \mid E_1, E_o) R_i(E_3 \mid E_2, E_1, E_o),$$

and any of the four other possible orderings of the three pieces of evidence gives the same overall likelihood ratio. Similarly, the weight-of-evidence formula gives the same result if any one or two of E_1, E_2, and E_3 are included with E_o. For example,

$$w_i(E_o) R_i(E_1, E_2, E_3 \mid E_o) = w_i(E_2, E_o) R_i(E_1, E_3 \mid E_2, E_o).$$

Senior UK judges have misunderstood this point in an important appeal case; see Section 9.3.4 below.

3.3.2 Assessing all the evidence

The weight-of-evidence formula (3.3) can be used to assess all the evidence in a criminal case. Suppose that the evidence can be allocated into four categories:

1. E_1 is information about the nature and location of the crime;
2. E_2 is an eyewitness description of the crime;
3. E_3 is the defendant's testimony;
4. E_d consists of the DNA evidence.

Initially, the jurors might assign values to the w_i on the basis of only E_1. After hearing E_2, the juror can calculate $R_i(E_2 \mid E_1) w_i(E_1)$. The probability of guilt $P(G \mid E_2, E_1)$ can be evaluated at this point, if desired. If E_3 is now taken into account, a new probability of guilt $P(G \mid E_3, E_2, E_1)$ can be calculated on the basis of the values of $R_i(E_3 \mid E_2, E_1) R_i(E_2 \mid E_1) w_i(E_1)$ for all i. These values can be regarded as the $w_i(E_o)$ for the purposes of assessing the DNA evidence E_d, where E_o stands for all the non-DNA evidence, E_1, E_2, and E_3.

Example

Suppose that there are 1000 possible culprits in addition to the defendant s, and the background evidence E_o does not distinguish among these 1001 individuals and so $w_i = 1$ for all i. Suppose that there are two items of evidence, E_{DNA} is a DNA match such that $R_i^{DNA} = 10^{-6}$ for all i, and E_{eye} is eyewitness evidence that may be assumed independent of the DNA evidence, so that the likelihood ratios multiply as in (3.7). Consider two possibilities:

- The eyewitness evidence supports the case against s, such that $R_i^{eye} = 1/100$ for every i. Then the probability of guilt is

$$P(G \mid E_{DNA}, E_{eye}, E_o) = \frac{1}{1 + \sum_{i=1}^{1000} w_i R_i^{DNA} R_i^{eye}}$$

$$= \frac{1}{1 + 1000 \times 10^{-6} \times 10^{-2}}$$

$$= \frac{1}{1 + 10^{-5}} \approx 0.99999$$

- The eyewitness evidence favours the innocence of s, such that $R_i^{eye} = 100$ for every i. Then the probability of guilt is

$$P(G \mid E_{DNA}, E_{eye}, E_o) = \frac{1}{1 + 1000 \times 10^{-6} \times 100}$$

$$= \frac{1}{1 + 10^{-1}} \approx 0.91$$

Thus, although much weaker than the DNA evidence, the eyewitness evidence can have a large effect on the overall conclusion.

3.3.3 The role of the expert witness

The weight-of-evidence formula requires modifications in some settings, such as when the crime sample has more than one source (Section 6.5). Nevertheless, the formula is very general and embodies the "in principle" solution to the problem of interpreting DNA profile evidence, including the role of the non-DNA evidence, the effects of relatives, population variability, and laboratory error. By "in principle", I mean that it points out the quantities that need to be assessed and how they should be combined.

One important feature of (3.3) in connection with DNA evidence is that it provides a demarcation of the roles of jurors and expert witnesses. Ultimately, it is for jurors to assess evidential weight, but (3.3) indicates that a DNA expert can be most helpful to clear-thinking jurors by guiding them with reasonable values for the R_i for various i, leaving the jurors to weigh these values together with the other evidence. Although there are usually many possible i, we have seen that it suffices to consider only a few cases. It will usually also be helpful for an expert witness to give some explanation of the logical framework for assessing evidence, so that the option is fully available to them. The w_i reflect jurors' assessments of the non-DNA evidence and will not usually be a matter for the (DNA expert) forensic scientist. The endorsement of any particular values for them should usually be avoided.

For most of this book, we will focus on applications of the weight-of-evidence formula to DNA evidence. We will assume that E_d refers to the DNA evidence, and that E_o includes all other evidence, so that we regard the DNA evidence as being assessed last. This is for convenience and is not necessary. A recommendation by a UK appeal court (Section 9.3.4) implies assessing the DNA evidence first, which has some advantages; in particular, it may then be reasonable to assume $w_i = 1$ for all i in $\mathcal{P}$.

For DNA evidence, whether other evidence such as information about alibis or eyewitness reports is included with the background information E_o will have little or no effect on the likelihood ratio $R_i(E_d \mid E_o)$. Note, however, that background information about the ethnic groups of i and s, or the relatedness of i with s, does affect likelihood ratios for DNA evidence. Any individual can make their own assessment of the probability $P(G \mid E, B)$ that the defendant is guilty, on the basis of the evidence E and any background information B that they feel appropriate. A juror's reasoning is, however, constrained by legal rules. For example, although it may be reasonable to believe that the fact that a person is on trial makes it more likely that they are guilty, a juror is prohibited from reasoning in this way (because otherwise evidence may be double-counted).

In the next section, we consider various consequences of the weight-of-evidence formula for assessing DNA evidence. We continue to assume that the R_i are given; we defer computing likelihood ratios until Chapter 6, after some necessary population genetics has been introduced in Chapter 5.

3.4 Consequences for DNA evidence

3.4.1 Many possible culprits

Because DNA evidence is widely, and correctly, perceived as being very powerful, both in convicting the guilty and in overturning past wrongful convictions (see e.g. Johnson and Williams 2004), cases often arise in which there is little or no evidence against the defendant other than the DNA evidence. In such cases, there may be large numbers of individuals who, if not for the DNA evidence, would be just as likely to be the culprit as the defendant (in other words, many individuals i for whom w_i is not small).

Many commentators seem to take the view that the fact that the profile is rare (i.e. p is small) alone establishes guilt. Some examples of statements, made by expert commentators who should have known better, that seem to be based on this fallacy are given below:

- "There is absolutely no need to come in with figures like 'one in a billion', 'one in ten thousand' is just as good";
- "The range may span one or two orders of magnitude, but such a range will have little practical impact on likelihood ratios as large as several million";
- "... population frequencies ... 10^{-5} or 10^{-7}. The distinction is irrelevant for courtroom use".

These statements are misleading because in the presence of many possible culprits, or strong exculpatory evidence, very small likelihood ratios may be consistent with acquittal, and differences of one or two orders of magnitude may be crucial.

The reason that very small likelihood ratios R_i may not suffice to imply a high probability for the defendant's guilt is that the bottom line (denominator) of the weight-of-evidence formula (3.3) involves a summation, and the total of many small quantities may not be small. A juror told only that 1 in 1 million persons has this profile may incorrectly conclude that this amounts to overwhelming proof of the defendant's guilt. This error can be extremely detrimental to defendants when there are many alternative possible culprits, or substantial exculpatory evidence.

3.4.2 Incorporating the non-DNA evidence

Consider two assault cases for which the non-DNA evidence, E_o, differ dramatically:

(i)
- the victim recognizes the alleged assailant and reports his name, s, to the police;
- s is found to have injuries consistent with the victim's allegation and cannot give a convincing alibi for his whereabouts at the time of the alleged offence;
- s is profiled and found to match the crime-scene profile.

A chance match

In 1999 a man from Swindon, UK, was found from the national DNA database to have a six-locus STR profile that matched the profile from a crime scene in Bolton, over 300 km away. Despite the distance, and the fact that the man was disabled, the reported match probability of 1 in 37 million was sufficiently convincing for the man to be arrested. A full ten-locus profile disclosed non-matches, and he was released the same day.

The press made much of this event, describing it as undermining the credibility of DNA evidence. The UK *Daily Mail* reported a senior criminal review official as saying

> "Everybody in the UK who has ever been convicted on six-[locus] profiling will want to apply to us to have their convictions reviewed".

The match probability was frequently misrepresented as, for example, in a local newspaper:

> "He was told that the chances of it being wrong were one in 37 million",

while *USA Today* said

> "... matched an innocent man to a burglary – a 1-in 37 million possibility that American experts call 'mind-blowing' ".

The incident was unfortunate for the man concerned, and perhaps the police can be criticized for not taking the distance and disability into account before making the arrest. However, the incident has no adverse implications for DNA evidence. Indeed, the match probability implies that we expect about two matches in the United Kingdom (population $\approx$ 60 million), and there could easily be three or four.

Because only a minority of individuals in the population currently has their DNA profile recorded, the extra matches would not be expected to be observed in any given case, but it is unsurprising that this eventually occurred. An analysis that takes all the alternative possible culprits into account can allow for unobserved, spurious matches.

ASSESSING EVIDENCE VIA LIKELIHOOD RATIOS

(ii)
- the victim does not see his assailant and can give no useful information about him;
- the crime profile is compared with DNA profiles from many other individuals until a matching individual s is found.
- s lives in another part of the country, has a good alibi for the time of the crime, and no additional evidence can be found linking him with the alleged offence.

The overall weight of evidence against the defendant, s, is very different in the two cases: in the first case, the evidence against s seems overwhelming; in the second, a jury would have to make careful judgements about the validity of the alibi, the possibility of travelling such a distance, and the strength of the DNA evidence.

The differences between these two cases is encapsulated in different values for the $w_i(E_o)$. A plausible allegation by the victim in case (i) may lead a juror to assign small values of w_i to each alternative possible culprit i. Lacking such an allegation, and faced with strong alibi evidence, jurors in case (ii) may assign values greater than one to many of the w_i. Of course, a juror may be reluctant to assign precise values to the w_i but can make broad distinctions between the moderately large and extremely small values appropriate in these two examples.

As noted above in Section 3.2, the w_i can be calculated using (3.3) sequentially, considering all the items of non-DNA evidence one at a time.

3.4.3 Relatives

Because DNA profiles are inherited, closely related individuals are more likely to share a DNA profile than are unrelated individuals. Many commentators have taken the view that close relatives of the defendant need not be considered unless there is specific evidence to cast suspicion on them.

The weight-of-evidence formula shows this view to be mistaken. Suppose that there is direct DNA profile evidence against a defendant s but the DNA profiles of the other possible culprits—a brother of s named b and 100 unrelated men—are not available. The non-DNA evidence does not distinguish between these 102 individuals, so that the "other evidence" ratios w_i are all equal to one. Let us assume the following values for the likelihood ratios: for the brother, $R_b = 1/120$ and for all other possible culprits $R_i = 1/1\,000\,000$. Then

$$P(G \mid E) = \frac{1}{1 + 1/120 + 100/1\,000\,000} \approx 99\%, \tag{3.8}$$

The probability that s is innocent is thus about 1%. A juror may or may not choose to convict on the basis of this calculation: the pertinent point is that ignoring the brother would give a very misleading view of evidential strength, leading to a probability of innocence of only 0·01%. It is easy to think of similar situations, involving additional unexcluded brothers or other close relatives, in which the probability of innocence is substantial, even after apparently strong DNA evidence has been taken into account. For an actual case in which the possibility that a brother was the culprit had an important effect on the outcome of an appeal, see Section 9.3.5.

It may be helpful to profile brothers in such cases, if possible. The brother may, however, be missing, or refuse to co-operate.[1] It may not even be known whether s has any brothers. Although still not realistic, the above example serves to illustrate the general point that consideration of unexcluded close relatives may be enough to raise reasonable doubt about the defendant's guilt even when there is no direct evidence to cast suspicion on the relatives.

3.4.4 Laboratory and handling errors

If crime and defendant profiles originate from the same individual, the observation of matching profiles is not surprising. Non-matching profiles could, nevertheless, have arisen through an error in the laboratory or at the crime scene, such as an incorrect sample label or laboratory record, a contaminated sample, a software error in a computer-driven laboratory procedure, or deliberate tampering with the evidence. The common practice of ignoring this possibility favours the defendant, although the effect is typically small.

On the other hand, there are at least two ways in which an observed match could have arisen even though s is not the culprit:

(i) suspect and culprit happen to have matching DNA profiles and no typing error occurred ("chance match");

(ii) suspect and culprit have distinct DNA profiles, and the observation of matching profiles is due to an error in one or both recorded profiles ("false match").

Both (i) and (ii) are typically unlikely. In many cases (ii) may be important but is rarely explicitly considered in conjunction with (i). Many arguments have been advanced for this (for counter-arguments, see Thompson *et al.* 2003), the most cogent of which is that error probabilities are difficult to assess. Even if error rates from external, blind trials are available, there will usually be specific details of the case at hand that differ from the circumstances under which the trials were conducted, and which make it more or less likely that an error has occurred.

We saw the role of false-match error probabilities under very simple assumptions in Sections 2.3.3, and further details will be given in Section 3.5.4. Some broad conclusions of these analyses are as follows:

- In order to achieve satisfactory convictions based primarily on DNA evidence, prosecutions need to persuade juries that the relevant false-match probabilities are small.

---

[1] In one case in which I was involved, the police, aware of this difficulty, arrested a brother of the principal suspect in order to compel him to give a DNA sample and hence exclude him from consideration as an alternative culprit. However, the arrest was ruled illegal by the court and the resulting DNA profile ruled inadmissible.

- The likelihood ratio R_i involves summing together the chance-match and the false-match probabilities.

- What matters is not the probability of *any* profiling or handling error but only the probability of an error that could have led to the observed DNA profile match.

It follows from the second point above that ignoring the possibility of false-match error is always detrimental to s, sometimes substantially so. If the false-match probability (ii) is judged to be much larger than the chance-match probability (i), then the latter probability is effectively irrelevant to evidential weight. Thompson *et al.* (2003) argue that jurors may mistakenly believe that a small false-match probability can be ignored, misunderstanding the point that it is not the absolute but the *relative* magnitude of the false-match to the chance-match probabilities that determines whether the former can be safely neglected.

Some critics of the use of DNA profiling often make the error of ignoring the final point. Dawid and Mortera (1998) point out that this error is similar to that highlighted in the eighteenth century by Price in response to an argument of Hume. Even if a printing error is more likely than winning the lottery, a newspaper report that your number has won should not therefore be dismissed: it is not the probability of *any* printing error that is pertinent but only that of an error that led to your number being reported, and this is typically less likely than an accurate report of your number. In contrast, with typical DNA evidence, a chance match is so extremely unlikely that the alternative possibility of the defendant's profile arising through error or fraud may well be much more likely, though still very unlikely in absolute terms.

3.4.5 Database searches

Many countries maintain national databases of the DNA profiles of named individuals for criminal intelligence purposes.[2] In the United Kingdom, the national database now records DNA profiles of well over 2 million individuals convicted of previous offences (in comparison, over 5 million fingerprint records are also held). These currently produce over 1000 matches per week with crime-scene DNA profiles, and about 40% of these "hits" lead to a resolution of the crime investigation.[3] Perhaps because of these successes, a recent extension of police powers in the United Kingdom to take samples from anyone arrested, whether they are charged or not, attracted little adverse comment despite its negative civil-liberties implications. In my view, a truly national database that includes the profiles of all citizens would be preferable to the present system of, in effect, police discretion as to who is included. For further discussion, see Kaye and Smith (2003).

---

[2] We use "population database" to designate a collection of the DNA profiles of unnamed individuals, and "intelligence database" when the profiles have names attached.

[3] For further details, see www.forensic.gov.uk.

The question thus arises as to the appropriate method for assessing the DNA profile evidence when the defendant was identified following a search through a database. The number of individuals involved in such a search, and even the fact that there was a search, may not be reported to the court. This is because intelligence databases consist primarily of the DNA profiles of previous offenders, and admitting that such a search has been conducted is thus tantamount to admitting previous convictions, and this is not permitted in many legal systems. It is important to know whether withholding this information from jurors tends to favour the prosecution.

In Section 2.3.4, we considered the related problem of a sequential search in the population of possible offenders in the setting of the island problem. As we discussed there, it is widely – but wrongly – believed that the fact that a DNA profile match is more likely when it results from a search means that the evidence is weakened by the search. In Section 6.1.1, we will discuss hypothesis formulations that lead to this erroneous conclusion. Many commentators, including the US National Research Council (1996), have made important errors in connection with intelligence databases (Section 9.4).

An analysis that focusses on the question of interest – is s the source of the crime-scene DNA profile? – shows that Lesson 5 (page 18) still holds, and DNA evidence is usually slightly stronger in the database search setting than when no search has occurred. This analysis does not require any modification of the weight-of-evidence formula. The evidence to be evaluated in R_i now includes all the DNA profiles observed, including all the profiles in the database that was searched. However, we argue in Section 6.2.6 that it is usually acceptable to ignore the non-matching profiles; doing so slightly favours s. The intuition behind this result is twofold:

(i) the other individuals in the database were found not to match and hence are effectively excluded from suspicion: R_i is negligible for all i whose profile is in the database but does not match the crime-scene profile;

(ii) the observation of many non-matches strengthens the belief that the profile is rare, so that each R_i should be (very slightly) reduced by the observed non-matches.

As an illustration of (i), imagine an enormous database that records the DNA profiles of everyone on earth. If the defendant's profile were the only one in this database to match the crime-scene profile, then the evidence against him would be overwhelming.

Although the DNA evidence may be (slightly) stronger in the context of a database search, the overall case against the defendant may be weaker because there is little or no non-DNA evidence against the defendant. For details of the analysis and discussion, see Balding and Donnelly (1996) and Balding (2002).

3.5 Some derivations †

So far I have stated many results without derivation. In this, more technical, section I fill in some of the missing details, which should be of interest to those who have mastered a first course in probability.

ASSESSING EVIDENCE VIA LIKELIHOOD RATIOS

3.5.1 Bayes theorem for identification evidence

Given evidence E and the two hypotheses confronting a criminal juror,

G: the defendant is guilty, and

I: the defendant is not guilty,

Bayes Theorem describes how to update *prior* probabilities of G and I to take into account the information conveyed by E. The Theorem is

$$P(G \mid E) = \frac{P(E \mid G)P(G)}{P(E \mid G)P(G) + P(E \mid I)P(I)} \quad (3.9)$$

All the probabilities in (3.9) are conditional on background information E_o, but we suppress this in the notation here. Since exactly one of G and I is true, we must have

$$P(G) + P(I) = 1 = P(G \mid E) + P(I \mid E).$$

Although valid, (3.9) is not immediately useful for DNA evidence because the likelihood $P(E \mid I)$ cannot be directly calculated. If the defendant is the source of the crime-scene DNA (which for now we assume is equivalent to guilt, see Section 6.1), then the probability $P(E \mid G)$ of observing the DNA evidence is relatively straightforward: it is the probability that an individual has a particular profile. If the defendant is not the source, however, we cannot evaluate the probability of the DNA evidence without knowing something about the person who was the source.

To overcome this problem, it is convenient to partition I into events of the form $C = i$, where C is the culprit and i denotes an individual other than the defendant. Then

$$P(I) = \sum_i P(C = i),$$

and

$$P(E \mid I)P(I) = \sum_i P(E \mid C = i)P(C = i).$$

Substituting in (3.9) and relabelling G as $C = s$ leads to the weight-of-evidence formula, (3.3).

Replacing I with $\cup_i \{C = i\}$ is convenient in many settings but is not appropriate in every forensic identification setting. One exception is the case that there is more than one unknown contributor to the crime-scene DNA profile (Section 6.5). In that case, we still cannot evaluate $P(E \mid C = i)$ even when we know that i was a contributor: we need also to know the other contributors in order to calculate the probability of the observed DNA profiles. See also the discussion of the choice of hypotheses in Section 6.1.

3.5.2 Uncertainty about p and N

Equation (2.2), which is the island problem formula modified to take into account uncertainty about p, can be derived by taking expectations in both the numerator and denominator of the likelihood ratio (3.4). Let the proportion of Υ-bearers in the island population now be $\tilde{p}$, a random variable with mean (i.e. average) p and variance σ^2. The denominator of R_i is the probability that a particular islander is a Υ-bearer. If the value of $\tilde{p}$ were known, this would be the required probability. Since $\tilde{p}$ is unknown, we must use its average, or expected value, $\mathrm{E}[\tilde{p}] = p$. Thus, the denominator of R_i is unaffected by the uncertainty. The numerator of R_i is the probability that two distinct individuals, s and i, are both Υ-bearers, which is the expectation of $\tilde{p}^2$, written $\mathrm{E}[\tilde{p}^2]$. From the definition of variance

$$\mathrm{Var}[\tilde{p}] = \mathrm{E}[\tilde{p}^2] - \mathrm{E}[\tilde{p}]^2,$$

it follows that

$$\mathrm{E}[\tilde{p}^2] = p^2 + \sigma^2,$$

and so (3.4) becomes,

$$R_i = \frac{p^2 + \sigma^2}{p} = p + \sigma^2/p.$$

As we noted in Section 2.3.1, the magnitude of the effect of uncertainty about p on the value of R_i can be very large: often σ^2/p is much larger than p. This is because of population genetic effects to be discussed in Chapter 5. Intuitively, the observation of a particular DNA profile makes it much more likely, because of common ancestry, that another copy of the same profile will also exist.

Whereas uncertainty about p affects the likelihood ratios R_i, uncertainty about N affects the prior probability, $P(G)$. Let the number of innocent islanders be $\tilde{N}$, a random variable with mean N. Conditional on the value of $\tilde{N}$, the prior probability of guilt is $P(G \mid \tilde{N}) = 1/(1 + \tilde{N})$, but since $\tilde{N}$ is unknown we need to use the expectation:

$$P(G) = \mathrm{E}[G \mid \tilde{N}] = \mathrm{E}\left[\frac{1}{1 + \tilde{N}}\right].$$

Because $1/(1 + \tilde{N})$ is not a symmetric function of $\tilde{N}$, but is convex (the function curve lies below the straight line between any two points on it), it follows from a fundamental result of probability theory (Jensen's inequality) that whatever the probability distribution for $\tilde{N}$, provided that $\mathrm{E}[\tilde{N}] = N$, the prior probability of guilt is never less than in the N known case. That is,

$$P(G) = \mathrm{E}\left[\frac{1}{1 + \tilde{N}}\right] \geq \frac{1}{1 + N}.$$

ASSESSING EVIDENCE VIA LIKELIHOOD RATIOS

The bound is tight unless the variance of $\tilde{N}$ is very large. Returning to the example of Section 2.3.2, if

$$\tilde{N} = \begin{cases} N-1 & \text{with probability} \quad \epsilon \\ N & \text{with probability} \quad 1-2\epsilon \\ N+1 & \text{with probability} \quad \epsilon, \end{cases}$$

then

$$\begin{aligned} E\left[\frac{1}{1+\tilde{N}}\right] &= \frac{\epsilon}{N} + \frac{1-2\epsilon}{1+N} + \frac{\epsilon}{2+N} \\ &= \frac{1}{1+N} + \frac{2\epsilon}{N(1+N)(2+N)} \\ &\geq \frac{1}{1+N}. \end{aligned} \quad (3.10)$$

Using (3.10) for the prior probability of G in the weight-of-evidence formula leads to the approximation (2.3) for the island-problem formula.

Because uncertainty about N does not affect the likelihood ratios, the higher prior probability of guilt implies a higher posterior probability than in the N-known case. Ignoring this uncertainty thus tends to favour defendants, though the effect is usually small. In practice, uncertainty about the population size is usually dealt with by replacing N with an upper bound, which also tends to favour defendants.

3.5.3 Grouping the alternative possible culprits

We noted in Section 3.2 that, for DNA identification evidence, it usually suffices to group together the possible culprits that have the same, or similar, degree of relatedness to s, and hence that have the same, or similar, value of R_i. If, for example, $\mathcal{P}$ is partitioned into three groups, a, b, and c, and the largest likelihood ratio for the individuals in each group is R_a, R_b, and R_c, respectively, then (3.3) becomes

$$P(G \mid E_d, E_o) \geq \frac{1}{1 + R_a \sum_{i \in a} w_i + R_b \sum_{i \in b} w_i + R_c \sum_{i \in c} w_i}. \quad (3.11)$$

If the likelihood ratios are nearly the same in each group, then the bound in (3.11) will be tight.

If the w_i are also the same over all i in each group, then we can write

$$P(G \mid E_d, E_o) \geq \frac{1}{1 + N_a R_a w_a + N_b R_b w_b + N_c R_c w_c}, \quad (3.12)$$

where N_a, N_b, and N_c are the numbers of individuals in each group. If these are unknown, then the comments concerning uncertainty about N in Section 3.5.2 apply to each of them.

3.5.4 Typing errors

Consider once again the modification to the island problem discussed in Section 2.3.3, in which typing errors occur independently with probabilities ϵ_1 and ϵ_2. Here, if suspect and culprit are not the same person, then the evidence must have arisen in one of three ways:

- Both suspect and culprit have Υ, and no typing error occurred; this has probability $p^2(1-\epsilon_2)^2$.
- One of the two has Υ, the other does not but a false positive error occurred; this has probability $2p(1-p)\epsilon_1(1-\epsilon_2)$.
- Neither suspect nor culprit has Υ, and both were incorrectly typed; this has probability $(1-p)^2\epsilon_1^2$.

If suspect and culprit are the same person, then there are two ways to have observed the evidence:

- The suspect/culprit has Υ and was correctly typed twice; this has probability $p(1-\epsilon_2)^2$.
- The suspect/culprit does not have Υ and was incorrectly typed twice; this has probability $(1-p)\epsilon_1^2$.

Combining all these probabilities, we obtain

$$R_i = \frac{p^2(1-\epsilon_2)^2 + 2p\epsilon_1(1-p)(1-\epsilon_2) + (1-p)^2\epsilon_1^2}{p(1-\epsilon_2)^2 + (1-p)\epsilon_1^2}$$

$$= \frac{(p+\epsilon_1 - p(\epsilon_1+\epsilon_2))^2}{p(1-\epsilon_2)^2 + (1-p)\epsilon_1^2}$$

$$\approx (p+\epsilon_1)^2/p,$$

The final approximation holds if p, ϵ_1 and ϵ_2 are all small.

3.6 Further weight-of-evidence exercises

Solutions start on page 160

1. Suppose that the evidence E_d is that glass fragments, whose refractive index matches that of a window broken during a crime, were found on the defendant's clothing. Assume here that there was a single perpetrator of the crime, the same person who broke the glass. Recall that according to our weight-of-evidence theory, we need to assess the likelihood ratio, R_i, which is the probability of observing E_d if some individual i committed the offence, divided by this probability if the defendant s did it (assume here that the effect of the other evidence E_o applies equally to all possible culprits, and so $w_i = 1$ for all i).

ASSESSING EVIDENCE VIA LIKELIHOOD RATIOS 41

(a) Think of a specific individual i. What does a rational juror need to assess in order to calculate R_i?

(b) How might the assessment described in (a) vary for different i?

(c) What information might an expert witness reasonably be able to provide to assist the juror in making these assessments?

2. This exercise is loosely based on the UK case R v. Watters, briefly outlined in Section 9.3.5. Suppose that there are 100 000 unrelated possible culprits in addition to the defendant s, and for each of these we have $w_i = 1$ and $R_i = 1/87$ million. Assume that for each of the two brothers of s we have $w_i = 1$ and $R_i = 1/267$ and that there are no other close relatives of s among the possible culprits.

(a) What is the probability that s is the culprit?

(b) How does the probability in (a) change if we now exclude the two brothers ($w_i = 0$ for each of them)?

3. This exercise is loosely based on the UK case R v. Adams (Section 9.3.3). The prosecution case rested on a DNA profile match linking the defendant with the crime, for which a match probability of 1 in 200 million was reported (the defence challenged this figure). The defence case was supported by an alibi witness who seems not to have been discredited at trial. The victim stated in court that the defendant s did not resemble the man who attacked her nor did he fit the description that she gave at the time of the offence.

(a) Consider only the DNA evidence, and assume that the 1 in 200 million figure is appropriate. It was reported in court that there were about 150 000 males between the ages of 15 and 60 within 15 km of the crime scene. On the basis of this information only, what might be a reasonable value for the probability that s is guilty? (Assume that these 150 000 males are all unrelated to the defendant.)

(b) Update this probability taking the alibi evidence into account, supposing that this evidence is 4 times more likely if s is innocent (irrespective of the true culprit) than if he is guilty.

(c) Now take the victim's evidence into account, supposing that this evidence is 10 times more likely if s is innocent (for all true culprits) than if he is guilty.

(d) Now suppose that the figure of 1 in 200 million is overstated, and that the correct value is 1 in 20 million. What approximate probability of guilt might you arrive at if you were a juror?

(e) Criticize the analysis of (a) through (d): what assumptions are unreasonable?

(f) How might you convey the implications of the above analysis to jurors? Alternatively, if you regard this as impossible or undesirable, give your reasons.

4. Return to the original island (Section 2.2.1) with 101 inhabitants each of whom has (independently) probability $p = 0.01$ of Υ-possession. If all 100 remaining islanders are untested, we calculated that the probability that s is guilty is 50%.

Now we are told that s was identified because he was found to be the only Υ-bearer among the 21 islanders whose Υ-status is recorded in the island's criminal intelligence database.

(a) Assume that inclusion in the database does not in itself affect the probability of guilt for this crime. What is the probability that s is guilty?

(b) Suppose that you are an observer at the trial, and not bound by the rules applying to jurors. You assess that the 21 individuals whose Υ-status is recorded in the database are each *a priori* 10 times more likely than any of the other islanders to have committed the crime. What should be your personal probability that s is the culprit, given the observation that he is the only Υ-bearer among these individuals?

5. Leaving the island again, consider a more realistic database search scenario in which most of the individuals in the database are not plausible alternative culprits for the crime, either because they were in jail at the time or because they live very far from the crime scene.

(a) Describe, in general terms, the effect on the evidential weight of a match found via a search in such a database.

(b) Consider the scenario outlined on page 32, in which an individual found to match in a database search was very unlikely to have committed the offence because he was ill at the time and was far from the crime scene. How, using our weight-of-evidence approach, would we assess the situation?

(c) If the unique matching individual in the database is definitively excluded from suspicion, what alternative lines of enquiry are suggested by the match?

4

Typing technologies

So far in this book we have been considering general principles for the evaluation of evidence, particularly rare-trait identification evidence. In this chapter, we begin the task of bringing these principles to bear on the DNA profile evidence in current forensic use. In this chapter, we briefly introduce the basic genetics and technology underlying DNA profile evidence, and in Chapter 5, we explore some relevant population genetics. Then we are ready to begin calculating likelihood ratios (Chapters 6 and 7) and presenting the evidence in court (Chapter 9).

A full account of the collection, management, and analysis of evidence is beyond the scope of the book, and indeed of the author's expertise. On the other hand, it is clear that error is always possible in any human endeavour, and that the assessment of the possibility of error is an essential part of the juror's task, and one that a scientific expert witness should seek to assist as far as possible. Thus, we need to consider the possible sources of error in different DNA profiling technologies and how these might affect evidence interpretation.

The most common sources of errors in connection with DNA evidence are routine handling and clerical errors. These are likely to be equally prevalent whatever the typing technology used, and may often be high. A recent European collaboration exercise to attempt to co-ordinate mtDNA typing (Parson *et al.* 2004) recorded that an error occurred in at least one participating lab for over 10% of the samples. Of these, 75% were clerical errors or sample mix-up, which, fortunately, are unlikely to lead to a false match.

Here, we focus on possible causes of error that are specific to different technologies. Currently, DNA profile evidence is predominantly in the form of short tandem repeat (STR) profiles obtained from unlinked, autosomal (i.e. not the mtDNA or X or Y chromosome) loci. We first give a brief introduction to the technology underlying STR profiles, and then introduce some other current and possible future types of DNA evidence. In each case, the coverage is not thorough but serves only to

Weight-of-evidence for Forensic DNA Profiles David Balding
© 2005 John Wiley & Sons, Ltd   ISBN: 0-470-86764-7

highlight specific issues relevant to interpretation (further references were given in Section 1.3). For comparison, we also briefly discuss fingerprint evidence.

4.1 STR typing

The basic steps in the production of an STR profile from a biological sample are:

1. extraction of DNA;

2. amplification via PCR (Polymerase Chain Reaction);

3. separation of PCR products according to length (strictly speaking the separation is according to electric charge, which has a close relationship with sequence length);

4. detection via staining or fluorescent dyes.

Step 1 will vary according to the source material, which may include hairs, biological fluids such as blood, semen, or saliva, or cellular material deposited on, for example, clothing, mobile phones, tools, or cigarette lighters.

Step 3 was originally achieved via a gel, a solid matrix with pores that permit DNA fragments to move in a buffer solution under the influence of the electric charge. After a fixed time interval, the solution is removed and the DNA fragments are immobilized in a solid gel. Today, this step is more often performed using capillary-based electrophoresis or slab-based gels, with detection via fluorescent dyes. Instead of PCR products being immobilized at locations that indicate their lengths, they all reach the end of the capillary or gel and the time taken is recorded: shorter PCR products reach the end faster than longer products. Included with the PCR products are size standards. When the fragments reach the end of the capillary, the PCR products and the size standards are excited by a laser and the resulting signals are recorded digitally, and graphically in the form of an electropherogram (EPG). The digital signals are processed to size the PCR products. The entire process of analysing fluorescence signals to obtain allelic designations is usually performed automatically by software supplied by the equipment manufacturer, although the allele calls are often reviewed visually from the EPG.

The "multiplexing" of 10 or more STR loci in a single electrophoresis run is achieved in two ways:

- **Length separation:** which is possible if all the alleles at one locus generate PCR products that are longer than all alleles at another locus;

- **Colour separation:** the signal from each dye is recorded on separate panels of the EPG.

Figure 4.1 shows an EPG together with allelic designations resulting from STR profiling of a buccal swab from a male individual, using ABI proprietary machines and software. Each of the four panels of the EPG shows the signal intensity over

TYPING TECHNOLOGIES

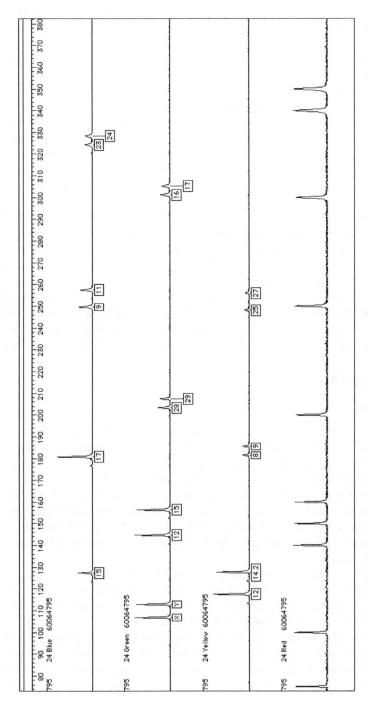

Figure 4.1 An STR profile of a male individual. The DNA sample was amplified with the SGM Plus™ STR kit. The amplified fragments were separated on the ABI Prism 377 DNA Sequencer and analysed using the GeneScan® 3.1.2 and Genotyper® 2.5 software. Image supplied courtesy of LGC. © 2004 LGC.

time for different colour dyes (all are shown here in black, but the dye colour is labelled on the diagram). The top ("Blue") panel records the individual's genotypes at four autosomal loci. The second ("Green") panel records the Amelogenin sex-distinguishing locus (leftmost; here recording an XY genotype for a male) and three autosomal loci. The third ("Yellow") panel records a further three autosomal loci, while the fourth ("Red") panel records size standards. The scale at the top of the diagram shows fragment length in base pairs (bp), with each notch corresponding to 2 bp. Thus, the two small peaks labelled 23 and 24 at the right end of the "Blue" panel are separated by 4 bp, as expected for consecutive alleles at a tetranucleotide STR locus. Size resolution to the nearest bp is usually feasible under good conditions. Note that the rightmost peaks on the EPG, corresponding to the longest alleles, tend to be lower than other peaks, and these are most susceptible to allelic "drop-out" when the quantity of DNA analysed is very small.

In Chapters 6 and 7 below, we will, for the most part, assume that the STR profiles have been recorded without error. For good-quality samples, such as that underlying Figure 4.1, this is usually a reasonable assumption: it is apparent from the figure that STR allele-calling is normally clear-cut. However, there are a number of anomalies that may arise even for good-quality samples. More importantly, crime-scene samples are often minute in quantity, and may also be contaminated and/or degraded, so that the above assumptions may be far from reasonable. We briefly discuss some anomalies that may arise in obtaining an STR profile. For further details, see Butler (2001) or Buckleton *et al.* (2004) and references therein.

4.1.1 Anomalies

Microvariants

STR alleles that include one or more repeat units that differ slightly from the prototype repeat motif are called *microvariants*. These can consist of a single-nucleotide insertion or deletion in one or more repeat elements, or may involve a sequence change that does not alter allele length. In the former case, the different alleles can usually be reliably distinguished. However, a microvariant that differs by 1 bp from the standard repeat length may be confused with *adenylation*: the addition of a non-template nucleotide, usually adenine, during PCR (see Butler 2001). Partial repeats are often rare, one of the exceptions being the THO1 allele 9.3, which is common.

Because electrophoresis in effect measures allele length, same-length microvariants are not distinguished: two PCR products of the same length will be recorded as matching alleles even if they differ at sequence level. This means that recorded STR profiles have less discriminatory capability than is potentially available via sequencing, but this causes no problem for the validity of the recorded match and its associated match probability. In particular, a defence claim that the profile of s may not match the crime-scene profile at the sequence level is true but irrelevant since the match probability incorporates all sequence variants of the same length.

Null alleles and allelic drop-out

Null alleles are STR alleles that, possibly because of a mutation in the primer sequence, are not amplified by PCR. A null allele can cause an individual who is truly heterozygous to be wrongly recorded as homozygous for the allele that is amplified. Individual null alleles are difficult to detect, but a high prevalence of null alleles at a locus may be detected via excess homozygosity (see Section 5.4.1). See Butler (2001) for the approaches used to minimize this problem, which is now rare with modern STR typing techniques.

Null alleles cause no problem for DNA profile interpretation provided that each null allele is consistently unamplified in repeat PCR assays. In that case, crime-scene and defendant profiles will correctly be recorded as matching if the defendant is the true source of the crime-scene DNA. This might be expected to occur if both crime scene and defendant samples were profiled in the same laboratory. Otherwise, differences in protocol, or use of PCR kits from different manufacturers, could generate a null allele in one laboratory that is non-null in another lab. As noted above, this rarely occurs and seems capable only of causing false exclusion errors rather than false inclusions. The false exclusion error rate can be minimized by retyping one or both samples if the profiles match at all but one or two alleles.

A related phenomenon is "allelic drop-out", in which one of the two alleles at a heterozygous locus is unamplified, or so weakly amplified that it cannot reliably be reported, because of unusual experimental conditions such as very low DNA copy number (Section 4.1.3). Because of possible allelic drop-out, minimum standards are usually imposed for interpreting a single peak as a homozygote genotype.

Stutter peaks

A "stutter peak" is a small non-allelic peak that appears on an EPG just before an allelic peak. It is caused by PCR products that have one (and sometimes two) repeat units less than the main allele. Left of centre in the "Blue" panel of Figure 4.1 is a high peak labelled 17, beside which is a very small, unlabelled peak at a location corresponding to an allele 4 bp shorter, that is, 16 repeat units. Since no other allele is detected at this locus, the possibility exists that the true underlying profile is a 16, 17 heterozygote, and that peak imbalance is manifest. However, such an extreme peak imbalance is very unlikely under normal conditions. In addition, the height of the peak at allele 17 supports the designation of the genotype as a 17, 17 homozygote at this locus, in which case the very small peak is dismissed as a stutter artefact. The height of the peak is the most important feature of the signal, but some other aspects, such as its shape (or *morphology*), and area beneath the curve, can also be informative in deciding on such designations.

Although stutter peaks could be a result of somatic mutation, it seems mainly to be an artefact of PCR, caused by mispairing of strands during replication. They usually have less than 10% of the height of the main peak, although the average height increases with allele length and can approach 15% for some very long alleles

(Butler 2001). Because of this, stutter peaks are usually easy to distinguish from real peaks if the DNA sample had a single contributor. However, for mixed crime-scene profiles (Section 6.5) for which the different sources contribute to the sample in very different proportions, the interpretation of stutter peaks can be problematic.

Stutter peak heights typically decrease with increasing length of the repeat unit. This is one reason that pentanucleotide STRs have been advocated to replace or supplement current STR loci in forensic use, which predominantly have a tetranucleotide repeat unit.

Over-stutter peaks are sometimes observed when large amounts of DNA are amplified by PCR. In this case, the spurious peak is at a location corresponding to one repeat unit greater than the allelic peak. The area of over-stutter peaks is usually at most 5% of that of the main peak. On rare occasions, they could be due to somatic mutations.

"Pull-up" peaks

These arise when an allele that is dyed with one colour also causes a peak to appear corresponding to one of the other dyes, which can be problematic if the location of the peak corresponds to an allele for each dye.

Mutations

STR mutations can be classified into two major types:

- **Germ-line (or meiotic) mutations:** these occur in the process of transmitting an allele from parent to child and can cause the child's allele to differ from its parental type.

- **Somatic (or mitotic) mutations:** these occur within an individual, in any cell of the body, and principally arise during cell duplication (mitosis).

Germ-line mutations can be important for paternity and other relatedness testing, see Section 7.1.8 below. For identification, it is somatic mutations that are potentially of some importance. Conceivably, a mutant type found in some but not all of an individual's cells could result in different profiles being recorded from different body tissue taken from the same individual, or perhaps from samples of the same tissue taken at different times. It is difficult to assess the chance of this occurring, but it is thought to be extremely rare. Somatic mutations seem capable of causing only false exclusion errors, not false inclusions.

Another possibility is that three distinct alleles could be recorded at a locus from the DNA of one individual. If this occurs in a crime-scene sample, it could create the mistaken impression that the sample contains the DNA of two individuals. Although potentially this could allow any individual with two of the three alleles at that locus to be wrongly included as a possible contributor, the phenomenon itself is very rare, and if it arises, the probability of a false inclusion, which requires a match at every other locus, is extremely small, and so this does not seem to be a realistic cause for concern.

TYPING TECHNOLOGIES

Gene and chromosome duplications

Some individuals have an entire extra chromosome ("trisomy"). This is usually fatal, except for trisomies involving chromosome 21 or the sex chromosomes. These often lead to severe disorders, although an extra X chromosome can be relatively benign. Gene duplication and triplication form a more realistic concern for forensic applications. Although still rare, copying of genes or other DNA segments to a new location on the same chromosome, or on another chromosome, seems to occur occasionally throughout the genome.

In these cases, one individual can have three or even four alleles amplified by a particular PCR primer. It may still happen that no more than two distinct alleles are represented, in which case the extra alleles will not usually be noticed although they can lead to peak imbalance. If a three-allele pattern does arise, it could wrongly suggest a mixed profile. However, if only one locus displays three peaks, and the peaks at the other loci are balanced, then gene duplication is a more likely explanation. Buckleton *et al.* (2004) briefly survey data suggesting a rate of triallelic profiles of between 1 and 20 per 10 000 single-locus genotypes (excluding sex-chromosome trisomies, which are more common).

4.1.2 Contamination

As techniques for collecting samples and extracting and amplifying DNA become ever more sensitive, the problem of contamination becomes potentially more serious. Our environment is to a large extent covered with DNA, expelled from its host by breathing or by touching. The effects of possible contamination have raised serious questions about the typing of ancient DNA and are similarly problematic for forensic work.

One precaution that is now often employed is for crime-scene and defendant profiles to be typed by different staff and in different laboratories (on the other hand, this can be problematic for ensuring that exactly the same protocols and interpretation standards are used for each). Other possible routine precautions include the use of ventilation hoods, wearing of face masks, and a policy of no talking during sensitive laboratory work. The DNA profiles of all police and scientific staff involved in investigating the crime scene or handling the evidence are usually kept for reference in case of contamination by one of these professionals.

Such precautions can greatly reduce, if not entirely eliminate, the possibility of contamination after evidence has been sealed. Nevertheless, since DNA evidence is often exposed to the environment for some time before being collected by crime investigators, the risk of contamination prior to, or during, collection cannot be ruled out. Contamination by "random" environmental DNA should be relatively harmless: it can complicate interpretation, but should not cause a false positive error. The most dangerous form of contamination is between different evidence samples at a crime scene, and possibly between crime scenes.

4.1.3 Low copy number (LCN) profiling

Low copy number ((LCN) refers to STR profiling from samples containing minute samples of DNA, possibly from just a few tens of cells. These may arise, for example, when DNA is extracted from fingernail debris, fingerprints, gloves, cigarettes, and so forth. There is no precise distinction between LCN and standard STR profiling: LCN techniques modify standard laboratory procedures to increase the sensitivity of STR profiling, for example, by increasing the number of PCR cycles, or reducing reaction volumes.

In principle, LCN profiling brings no new interpretation issues, but in practice, the sensitivity of the technique means that anomalies such as extreme imbalance of the peaks corresponding to the two alleles at a locus, the presence of contaminant alleles, as well as stutter peaks and other artefacts, are much more prevalent than in standard settings. One consequence is that it is frequently necessary to record a genotype as ambiguous: for example, if one allele is observed, it may be unclear whether this represents a homozygote or a heterozygote subject to allelic drop-out. For a discussion of the formulation of hypotheses and the calculation of likelihood ratios in the presence of such anomalies, see Gill (2001a), Evett *et al.* (2002), and Buckleton *et al.* (2004).

4.2 mtDNA typing

The mtDNA is a circular molecule of about 16·5 kilobases (Kb) in length – minuscule in comparison with an autosomal chromosome, which is typically around 10^5 Kb in humans. Unlike nuclear DNA, mtDNA resides outside the nucleus of the cell, and is carried by cytoplasmic energy-generating organelles called mitochondria. Whereas the nuclear genome exists in two copies per cell, one paternal and one maternal in origin, mtDNA exists in multiple copies – the copy number varies widely by cell type, depending on the cell's energy requirements.

Although it seems likely that paternal inheritance can occur in some rare situations, mtDNA is almost entirely maternally inherited (Birky 2001). If bi-parental inheritance does occasionally occur, then the possibility for recombination also arises. In the past few years, some evidence has been reported for recombination in mtDNA, but this has been disputed, and the present consensus seems to be that if mtDNA recombination does occur it is too rare to be important for most purposes (see e.g. Wiuf 2001).

mtDNA samples have been widely used to infer aspects of human female population histories (reviewed in Jobling *et al.* 2004). One of the key advantages of mtDNA over nuclear DNA for such work is its higher mutation rate, which generates substantial diversity. Also, because it exists in multiple copies, mtDNA is easier to type from small and/or degraded samples. In recent years, ancient human mtDNA has been successfully typed, although there remains controversy about the validity of the results and their interpretation (Jobling *et al.* 2004).

For the same reasons of genetic diversity and relative ease of typing from degraded samples, mtDNA profiles are also useful in forensic identification (Tully et al. 2001). They are widely used for samples containing little or no nuclear DNA, for example, shed hair, bone, and burnt or other seriously degraded remains. In addition, mtDNA has been successfully typed from DNA-poor sources such as saliva on stamps and tooth root dentin.

mtDNA is relatively gene rich (93% is coding sequence), but the "control region" of approximately 1·1 Kb is non-coding and is the focus of most interest for both population genetic and forensic work. Two portions of the control region, called hypervariable regions I and II, each of about 300 bp, have the greatest variability. This variability manifests itself in sequence alterations as well as in STR and poly-C length polymorphisms. mtDNA is the only DNA type for which sequencing is currently the usual method of profiling: an mtDNA "profile" typically consists of the entire sequence of one or both of the hypervariable regions, although the exact region sequenced seems not to be standardized between laboratories. See Rudin and Inman (2002) and Tully et al. (2001) for further discussion of mtDNA profiling technologies and interpretation standards. Recently Vallone et al. (2004) have proposed a set of 11 SNPs distributed throughout the mtDNA genome, for distinguishing individuals that match in the sequenced hypervariable regions.

One consequence of the high mtDNA mutation rate is heteroplasmy: the existence within an individual of multiple mtDNA types (differing at one or several sites). This is essentially the same phenomenon as the autosomal somatic mutations discussed above in Section 4.1.1, but mtDNA heteroplasmy occurs more frequently and is often associated with disease. It can be inherited, in which case it affects all the cells of the body in the same way.

Provided that mtDNA heteroplasmy can be reliably identified, it increases the discriminating power of mtDNA profiles. However, different typing procedures have different abilities to detect and record heteroplasmy and this could potentially lead to false exclusion errors. See Tully et al. (2001) for a further discussion of heteroplasmy and its effects.

4.3 Y-chromosome markers

The Y chromosome is relatively short for a nuclear chromosome (60 000 Kb) but still much longer than the mtDNA chromosome. For the most part, it is inherited uni-parentally, through the paternal line, although short, terminal sections of the Y recombine with the X chromosome (these "pseudo-autosomal" regions of the Y are not discussed further here). Unlike mtDNA, which is carried by both sexes, Y chromosomes are carried only by males. Also in contrast with mtDNA, the Y chromosome is gene poor, but it does carry the sex-determining SRY gene.

Overall, human Y chromosomes are relatively homogeneous, leading to estimates of a relatively short time (Wilson et al. 2003, report estimates < 50 000 years) since the most recent common ancestor of Y chromosomes. Further, the variation that does exist is predominantly between populations; Y chromosomes are relatively homogeneous within populations. This pattern of variation could be due to the effects of selection and/or a high between-male variation in reproductive success.

Some particular uses of Y-chromosome profiles in forensic work include identifying the male contributor to a mixture when the major component is from a female, or when the male donor is aspermic and so sperm-separation techniques cannot be employed. Y-chromosome profiles are also useful in distinguishing multiple male contributors to a mixture. As for the autosomes, STR markers form the predominant marker type for Y-chromosome forensic work, although SNP and other polymorphisms are available.

Information about Y-chromosome STR haplotype population frequencies is available at www.ystr.org. In rare cases, a supposedly Y-specific primer may amplify a homologous X sequence. In addition, duplication and triplication polymorphisms are relatively common on the Y chromosome, so that multiple alleles may occasionally be observed.

4.4 X-chromosome markers †

The X chromosome is, like the Y, involved in sex determination: girls receive an X from their father, boys receive a Y; both receive an X from their mother. Thus, unlike the Y, the X is not sex specific. At 165 000 Kb, the X chromosome is much longer than the Y. Although still relatively gene poor compared with the autosomes, it has a much higher gene density than the Y.

Many diseases are caused by a defective allele and are hence recessive in autosomal form (no adverse effect when paired with a normal allele). However, males inheriting a defective X allele are usually adversely affected, whereas females have the same protection from recessive diseases on the X as for the autosomes. This distinctive mode of inheritance of diseases is referred to as "sex linked". The most common sex-linked diseases are red-green colour blindness (which affects 8% of men but only 0·5% of women), hemophilia A, and several forms of mental retardation, including fragile-X syndrome.

X-chromosome STR markers are not as widely used in forensic applications as those described above, but they do have some distinct advantages in paternity testing of girls. Because a man has only one allele to transmit to his daughter, there is never ambiguity about the father's transmitted allele, so that X-chromosome markers are, on average, slightly more informative than autosomal markers. More importantly, if the girl's father is absent but her mother is available, the father's X-chromosome can be inferred and compared with the X chromosomes of close relatives of a putative father.

Because they are not yet widely used, we will not discuss X-chromosome markers any further in this book. See Szibor *et al.* (2003) for a discussion of their forensic uses and some basic statistical analyses, and Ayres and Powley (2004) for inclusion probability and paternity index calculations.

4.5 SNP profiles

The single-nucleotide polymorphism (SNP, pronounced "snip") has recently become the marker of choice for many human genetic studies, because:

- it is the most abundant type of polymorphism in the human genome;
- SNPs can be typed accurately, cheaply, and in large volumes.

SNPs are not necessarily diallelic, but this is predominantly the case, and almost all SNPs selected for typing are diallelic. In that case, it is meaningful to speak of the "minor" allele. SNPs with a relatively common minor allele (say, relative frequency >10%) are the most informative and are usually preferentially chosen for typing over SNPs with a rare minor allele.

It is natural to consider whether SNPs might make a more suitable typing system for forensic purposes than the STR markers that dominate current forensic work. Although a typical STR locus is as informative as 4 to 6 SNPs, the two advantages above can outweigh this disadvantage. A bank of 50 SNPs will typically have at least the same information content as a dozen STR markers. Much development work is currently in train, and comparative costs and benefits of SNP and STR typing are not yet fully clear. One particular advantage of SNPs is that they avoid the problem of stutter peaks. A disadvantage is that, necessarily, several SNPs will lie on the same chromosome. If sufficiently far apart, this linkage will cause little problem for identification, but it can be problematic for the analysis of relatedness in large pedigrees (Section 7.1.4).

Despite their actual and potential advantages, a recent working party concluded that SNPs were unlikely to replace STRs as the predominant marker system for forensic applications (Gill *et al.* 2004). The disadvantages cited include:

- Huge investment has already been made in accumulating large STR databases, for example, of previous offenders and of unsolved crimes. Retyping with SNPs would be expensive even if the samples had been kept to make it feasible.

- Although SNPs can be cheap to type in large volumes, for example, in the setting of genetic disease studies, the cost savings can be difficult to realize in forensic work, which can involve small and degraded samples requiring individual attention.

- SNPs are relatively poorly informative about contributors to mixtures, since at each SNP a heterozygote genotype will be recorded unless all contributors to the mixture are homozygous for the same allele. See Gill (2001b) for a further discussion.

- Along with STRs, SNPs suffer from problems of allelic drop-out in very small samples.

Although these authors argue that SNPs are unlikely to rapidly become the dominant forensic marker, they are already being used for some specialist applications – mtDNA and Y-chromosome typing and to predict some aspects of phenotype (Section 7.4) – and this usage seems likely to increase. SNPs also have advantages in analysing highly degraded samples that can arise from aircraft crashes and other disasters.

4.6 Fingerprints †

We are mainly concerned in this book with identification evidence that is directly measured from DNA sequences, but it seems useful to make some comments on, and comparisons with, fingerprint evidence, which in some respects is similar.

Although fingerprints are in part genetically determined, and hence can be regarded as an indirect measure of DNA variation, there is sufficient environmental variation in the process of foetal development that even identical twins have distinguishable fingerprints. This weaker correlation between relatives gives fingerprints an advantage over DNA profile evidence and is the basis of the widely cited claim that all fingerprints are unique.

It is impossible to prove any human characteristic to be distinct in each individual without checking every individual, which has not been done. The complexity of the variation makes the claim at least plausible, and it is further supported by the successful use of fingerprint evidence for over a century. Yet, although there have been a number of theoretical investigations, the suggestion that recorded fingerprints are unique has never been rigorously checked. See Kaye (2003) for a critique of a recent attempt to try to provide scientific back-up for the claim.

Whatever the truth of the uniqueness claim for true fingerprints, it is of little practical relevance. What is at issue in a court case involving fingerprint identification is not

"Is this person's (true) fingerprint unique?".

Given two people, it is probably true that an expert could always distinguish their fingerprints given careful enough measurements. Instead, in a practical identification scenario, the question is more likely to be

"Is this imperfect, possibly smudged, smeared, or contaminated, impression of a fingerprint taken from a crime scene enough to establish that the defendant and nobody else could have left it?"

The answer to that question will always depend on the quality of the impression recorded at the crime scene, even if the general proposition of the uniqueness of (true) fingerprints is accepted.

To help answer this question, UK police in 1924 adopted a "16 matching points" standard for reporting fingerprint evidence, with 10 matching ridge characteristics sufficing for subsequent marks at the same scene, provided that one of them reached the "16-point" standard. This standard was accepted for over 60 years, but has subsequently been demolished as having little scientific basis.

No specific alternative numerical standard has replaced it. In this respect, DNA evidence is far superior to fingerprint evidence, since uncertainty can be measured reasonably well, and given enough markers, it is extremely small in standard cases. In fact, DNA profile evidence is now seen as setting a standard for rigorous quantification of evidential weight that forensic scientists using other evidence types should seek to emulate.

5

Some population genetics for DNA evidence

5.1 A brief overview

5.1.1 Drift

If we sample 10 balls at random from a bag on several occasions, the number of green balls chosen may vary even though the contents of the bag have not changed. This is the chance variation due to sampling, sometimes called "sampling error". It is what the "±3%" that often accompanies opinion poll results is trying to measure.

The relative frequencies of alleles (gene variants) also differ from population to population, but the variation is often larger than that due to sampling error. In fact, the evolution of genes from generation to generation in a closed population is like repeated, *compounded*, sampling of balls from a bag. Suppose that there are initially 10 individuals in the population, 5 "green" and 5 "blue". Sex is an unnecessary complication for simple population genetics models, and we will assume that all 10 individuals are equally fit and are capable of morphogenesis to produce a large number of offspring the same colour as themselves. However, the environment will only support 10 adults, and so only 10 of the offspring survive to reproduce. In the absence of selective advantage, the number of green individuals in the second generation will have a random distribution centred at 5 (in fact a Binomial(10, 0·5) distribution, with mean 5 and standard deviation $\sqrt{10 \times 0·5 \times 0·5} \approx 1·6$).

Suppose that the actual number of green individuals in the second generation happens to be 7. The random process of reproduction occurs again, and the number of green individuals in the third generation has a Binomial(10, 0·7) distribution, with mean 7 and standard deviation $\sqrt{10 \times 0·7 \times 0·3} \approx 1·4$. The number of green individuals in the population will "drift" away from its starting value,

---

Weight-of-evidence for Forensic DNA Profiles   David Balding
© 2005 John Wiley & Sons, Ltd   ISBN: 0-470-86764-7

SOME POPULATION GENETICS FOR DNA EVIDENCE

until eventually "fixation" of greenness or blueness occurs: either everyone in the population is green or everyone is blue.

If there are many populations evolving in the same way and with the same starting configuration, then eventually about half of the populations will be all-green, and about half will be all-blue. Figure 5.1(a) shows the counts of green individuals

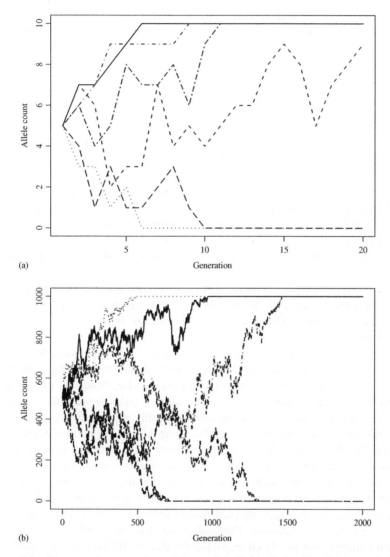

Figure 5.1 Frequencies over time of "green" individuals in 6 haploid populations. (a) Population size = 10, initial frequency = 5, number of generations simulated = 20. (b) Population size = 1000, initial frequency = 500, number of generations simulated = 2000.

in 6 populations, each evolved for 20 generations from a starting configuration of 5 green and 5 blue individuals. Five of the populations reached fixation, of which 3 were fixed for the green allele.

Alleles that have reached fixation are of little interest for forensic identification. In real human populations, there are many factors that tend to counter, or modify, the effects of drift. First, we consider the effects of population size, and then move on to consider other evolutionary processes: mutation, migration, and selection. We do not focus here on either recombination or gene conversion which, though important evolutionary processes, do not play an important role in the forensic issues discussed in this book.

Population size

Human populations are usually much larger than 10 individuals. Fortunately for us, under simple models, genetic drift behaves in approximately the same way for different population sizes, but at a speed that is inversely proportional to population size. Figure 5.1(b) illustrates a scenario in which the allele frequency and population size are both 100 times their values in (a) and the pattern of fixation is similar except that it occurs 100 times more slowly. For this reason, much of population genetics theory works with time scaled according to the population size: this way the same formulas can be applied to a population of any size.

The situation is more complicated because it is not the census population size that matters but the *effective* population size, which is usually much smaller and is difficult to measure. The effective population size is the size of a hypothetical random-mating population that displays the same level of genetic variation as the observed population: it is the size of the theoretical population that best fits the actual population. There are several ways to define "best fitting", and hence several different definitions of effective population size. Note that there may be no random-mating population that fits the actual population well: the concept of effective population size is useful but not universally applicable.

Surprisingly, the effective size of the entire human population is often estimated to be about 10 000 adults, many orders of magnitude less than the current census size of several billions. There are many possible reasons for this extremely low value; some among them are the effects of rapid population growth, geographic dispersal, locally fluctuating population sizes, high between-male variance in reproductive success, and inherited female fecundity. See Jobling *et al.* (2004) for a more extensive discussion. Remember that effective population size is a theoretical concept, and an effective population size of 10 000 does not imply that an ancestral population of size 10 000 had any special role in human history. It does mean that the actual human genetic variation looks in some respects like the variation that would be expected at neutral loci in a random-mating population of constant size 10 000.

5.1.2 Mutation

Mutation is the primary generator of genetic variation. Since drift tends to reduce genetic variation, an equilibrium can arise under neutral population-genetics models in which the genome-wide average amount of variation is maintained at the level at which the new variants being created by mutation are balanced by the variants being lost through drift. Possibly, the most important formula in population genetics is

$$F = \frac{1}{1 + 4N\mu}, \qquad (5.1)$$

in which F denotes the probability that two chromosomes are identical by descent (ibd) at a locus with mutation rate μ, under mutation-drift equilibrium in an unstructured, random-mating population of N diploid individuals. Here, ibd implies that neither chromosome has suffered a mutation at the locus since their most recent common ancestor.

Most of the human genome has a very low mutation rate: for a typical nucleotide site, $\mu \approx 2 \times 10^{-8}$, or about 2 mutations per 100 million generations, and consequently, at mutation-drift equilibrium, there is relatively little variation. Indeed, the probability that two homologous human chromosomes chosen at random (worldwide) match at a given nucleotide site is typically about 0.9992. Although a match does not imply ibd, because of the low mutation rate, the two are expected to be almost equivalent in practice, and substituting $F = 0.9992$ and $\mu = 2 \times 10^{-8}$ into (5.1) leads to $N \approx 10\,000$, as discussed above in Section 5.1.1.

The control region of mitochondrial DNA (Section 4.2) has a higher mutation rate, on average perhaps two orders of magnitude higher than autosomal nucleotides.

STR mutation rates are even higher (about 1 or 2 mutations per locus per thousand generations). Perhaps the most common theoretical model for STR mutation is the stepwise mutation model (SMM) in which a mutant allele has either $k - 1$ or $k + 1$ repeat units, each with probability 1/2, where k is the current repeat number. The SMM has no stationary distribution, so two similar populations isolated from each other do not converge under the SMM to the same allele frequency distribution. Thus, substantial between-population diversity is expected under the SMM for populations that exchange few migrants. In contrast, there is typically little between-population variation at human STR loci, suggesting high migration rates and/or the invalidity of the SMM.

In fact, the strict SMM is known to be false, for example, because the mutation rate increases with allele length and occasional two-step mutations occur, although it may provide an adequate approximation for some purposes. The SMM can easily be modified, for example, by hypothesizing a bias towards contraction mutations in long alleles, to obtain STR mutation models that do have a stationary distribution (Calabrese and Durrett 2003; Lai and Sun 2003; Whittaker *et al.* 2003; Xu *et al.* 2000). Such models can provide a better fit to observed data than the SMM and are consistent with the observed between-population homogeneity of allele proportions at many STR loci.

5.1.3 Migration

From a global perspective, migration does not generate new variation, but human populations are geographically structured and, from a local perspective, new migrants can be the most important source of genetic variation. Indeed, in simple models of a large population divided into subpopulations that exchange migrants, the formula describing the genetic diversity at a locus under migration-drift equilibrium exactly mimics the mutation-drift formula (5.1):

$$F = \frac{1}{1 + 4Nm}, \tag{5.2}$$

in which m is the proportion of subpopulation i that migrates to subpopulation j each generation, assumed to be the same for all i and j and constant over time. Here, F is the probability that the two chromosomes have a most recent common ancestor within the subpopulation, without any intervening migration event.

The important implication of (5.2) is that relatively low levels of migration (but still much higher than the mutation rate, which is neglected) can suffice for F to be small, allowing much higher genetic variation within subpopulations than would be expected in an isolated population subject only to mutation and drift.

For most of the human genome, mutation rates are very low. Moreover, substantial migration is known to be a recurrent phenomenon throughout human history and much of prehistory (Jobling *et al.* 2004). Thus, migration is thought to be the dominant factor in explaining the geographic distribution of human genetic variation. Indeed, genetic polymorphisms are used to trace ancient migrations (see e.g. Romualdi *et al.* 2002) and the effects of mutation often play little role in this work. At STR loci, the much higher mutation rate makes tracing migration events more difficult, other than on the Y chromosome when multiple completely linked STR markers can be exploited. Indeed, Rosenberg *et al.* (2002) found that nearly 95% of human genetic variation at autosomal STR loci was accounted for by within-population differences, and so very many STR loci are required to distinguish human populations.

If the STR mutation process is at least approximately stationary, which is supported by the recent data and models cited above in Section 5.1.2, then both migration and mutation tend to homogenize allele frequency distributions in different populations. There is a distinction in their effects in that a mutant STR allele tends to be similar in length to its parental type, whereas a migrant allele can, in principle, be of any allelic type. However, geographical structuring of human populations may mean that migrant alleles are also of a similar type to existing alleles in the subpopulation. Consequently, the effects of mutation are difficult to distinguish from those of migration in explaining current genetic variation.

5.1.4 Selection

Selection refers to the differential success of individuals according to their genetic make-up, where success is measured in terms of producing offspring and raising them to maturity. Some forms of selection can tend to reduce genetic variation within a (sub)population, such as the elimination of deleterious variants or the fixation of advantageous variants. Other processes, referred to collectively as balancing selection, tend to maintain genetic diversity, in some cases at approximately the same level in different populations that are isolated from each other. The most important forms of balancing selection are:

> **Heterozygote advantage:** individuals with two different alleles at a locus enjoy an advantage over homozygotes (this form of selection seems to be common in humans at loci involved in immune response); and

> **Frequency-dependent** selection; for example, an individual of a type that is rare in the population has an advantage over those of common types (for example, because it tends to escape the attention of predators).

A major controversy in late twentieth century population genetics was the extent to which observed genetic variation reflects the effects of selection versus those of drift. For a century after Darwin, it was widely accepted that most genetic variation had a selective explanation. The emergence of molecular variation data from the 1970s led to the neutral theory (see e.g. Kimura 1977). Although hotly contested, the neutral theory has had a profound impact so that, for the past one or two decades, most population geneticists have assumed that most human genetic variation can be regarded as approximately neutral for most purposes. The huge amount of genomic data now becoming available will allow this assumption to be checked more thoroughly.

Some STR loci are directly implicated in causing human disease: an elongated coding-sequence CAG repeat causes Huntington disease, and other trinucleotide repeats peripheral to coding sequences can interrupt transcription, causing, for example, fragile-X syndrome. The tetranucleotide repeats typical in forensic applications are generally thought to have little or no phenotypic effect and consequently not to be under direct selection. However, there is little substantial evidence to support this, and some evidence to the contrary. Albanese *et al.* (2002) reported an intronic tetranucleotide STR influencing transcription, and Subramanian *et al.* (2003) suggested a possible functional role for GATA repeats. Further, some long-established STR loci in forensic use are closely linked with disease-implicated genes (see Butler 2001, and references therein), and hence, the patterns of variation at these loci are potentially affected indirectly by selection.

The selectionist-neutralist debate is not fully resolved, but it is clear that (effective) population size is crucial: the effects of drift are relatively more important when the population size is small. The low human effective population size means that humans have less genetic variation overall, and are more susceptible to the effects of drift, than might have been expected given our large census population size. In the sequel, we will usually assume neutrality at the loci under consideration, and occasionally comment on the possible effects of selection.

Selection almost certainly plays an important role in explaining the population variation at mtDNA and Y-chromosome markers (Sections 4.2 and 4.3), since for these chromosomes there is no recombination to limit the extent of selective effects. However, because few population-genetic assumptions are made in the interpretation of these markers, it may have little effect on their role for forensic work (outlined in Section 6.4).

5.2 θ, or F_{ST}

Other factors, beyond those considered above, can modify the effects of drift, including admixture (e.g. when two previously distinct populations merge) and inbreeding. A detailed understanding of the influence of all factors on the evolution of profile proportions in human populations requires a lifetime of study, and more. Although some understanding of the underlying processes is helpful, for the purposes of forensic applications, the most important question concerns the *magnitude* of the variation of allele proportions among different subpopulations, relative to the population from which a forensic database was drawn.

For example, the population may correspond to "UK Caucasians", and the subpopulations relevant to a particular crime may include, for example, people of Irish, Cornish, East Anglian, Orkney Island, Cypriot, or Jewish ancestry. Because of migrations and intermarriages, such groups are not strictly well defined, and simple population-genetics models do not apply exactly. However, such models can form the basis for an analysis that attempts to assess the variation of allele proportions in such subgroups about a Caucasian average. The effects of this variation can then be allowed for in DNA profile match probability calculations, without a full knowledge of either the history of these groups or the population genetic forces that caused the variation.

Wright (1951) introduced the coefficient F_{ST}, more often called θ in forensic applications, which he interpreted as measuring the average progress of subpopulations towards fixation, and hence he called it a fixation index. $\theta = 1$ implies that all subpopulations have reached fixation at the locus, possibly for different alleles in different subpopulations; $\theta = 0$ implies that allele proportions are the same in all subpopulations, and so the population is homogeneous.

θ can also be interpreted in terms of the mean square error (MSE) of subpopulation allele proportions about a given reference value. If $\tilde{p}$ denotes the subpopulation proportion of an allele, and the reference value is p, then the MSE is the expected (or mean or average) value of $(\tilde{p} - p)^2$, and we have

$$\text{MSE}[\tilde{p}, p] = \text{E}[(\tilde{p} - p)^2] = \theta p(1 - p). \tag{5.3}$$

In forensic applications, the reference value is typically the allele proportion in a large, heterogeneous population from which a database has been drawn. The implications of this are discussed further in Section 6.3.2.

Usually in population genetics, the reference value p is the mean (either the average over all the populations, or the expected value under some evolutionary model), in which case the MSE is the same as the variance. For example, if $p = 0.2$ and $\theta = 1\%$, then the subpopulation allele proportions have mean 0.2 and standard

SOME POPULATION GENETICS FOR DNA EVIDENCE

deviation $\sqrt{0{\cdot}01 \times 0{\cdot}2 \times 0{\cdot}8} = 0{\cdot}04$. The statistical rule-of-thumb "plus or minus two standard deviations" then gives a rough 95% interval of 0·12 to 0·28 for a subpopulation proportion. This rule of thumb does not work well for p small, and we can compute better intervals using the beta distribution, introduced in Section 5.3.1 below. Here, the beta distribution gives almost the same 95% interval: (0·13, 0·28).

If $\tilde{p}_1, \tilde{p}_2, \ldots, \tilde{p}_k$ denote estimates of $\tilde{p}$ in k different subpopulations and they each have (known) expectation p and the same variance, then a natural estimator of this variance is

$$\text{Var}[\tilde{p}] \approx \frac{1}{k}\sum_{j=1}^{k}(\tilde{p}_j - p)^2.$$

Substituting in (5.3) leads to the estimator

$$\widehat{\theta} = \frac{\sum_{j=1}^{k}(\tilde{p}_j - p)^2}{kp(1-p)}, \tag{5.4}$$

in which we introduce the notation $\widehat{}$ to denote an estimator. For example, if $p = 0{\cdot}1$ and the allele proportions in three subpopulations are 0·05, 0·11, and 0·13, then using (5.4) we would estimate

$$\widehat{\theta} = \frac{(0{\cdot}05 - 0{\cdot}1)^2 + (0{\cdot}11 - 0{\cdot}1)^2 + (0{\cdot}13 - 0{\cdot}1)^2}{3 \times 0{\cdot}1 \times 0{\cdot}9} \approx 1{\cdot}3\%.$$

Usually, p is not known exactly and must be replaced with an estimate in (5.4). Often, the estimate of p is the value that minimizes $\widehat{\theta}$, and it is customary to replace the $1/k$ with $1/k-1$ in (5.4) to compensate for the bias that arises from this choice. Here, if p were unknown, the natural estimate would be $(0{\cdot}05 + 0{\cdot}11 + 0{\cdot}13)/3 = 0{\cdot}097$, which is close to 0·1 but leads to a much larger estimate $\widehat{\theta} \approx 2\%$ because the bias correction has a large effect when k is small. In forensic settings, the estimate of p will usually be based on the allele proportion in the most relevant population database.

There are a number of problems with using (5.4) in practice; among them is the fact that it assumes a common value of θ over subpopulations. For more sophisticated method-of-moments estimators, see Weir and Hill (2002), and for likelihood-based estimation, see Section 5.7.

5.3 A statistical model and sampling formula

5.3.1 Diallelic loci

The beta distribution

Between-population variation in allele proportions at a diallelic locus is often modelled by the *beta* distribution, which has probability density function[1] (pdf):

$$f(x) = cx^{\lambda p - 1}(1-x)^{\lambda(1-p)-1}, \tag{5.5}$$

---

[1] Our parametrization of the beta is not standard, but is convenient here. The standard parametrization has $\alpha = \lambda p$ and $\beta = \lambda(1-p)$.

where c is a normalizing constant whose value is known but not needed here, $0 \leq x \leq 1$, and

$$\lambda = \frac{1}{\theta} - 1.$$

The expectation and variance of the beta are respectively p and $\theta p(1-p)$.

Plots of the beta distribution for various p and $\theta = 1\%$, 2%, and 5% are shown in Figure 5.2. Each curve represents a theoretical distribution of the proportion of the "green" allele in a subpopulation. For example, the solid curve in (c) ($p = 20\%$, $\theta = 1\%$) indicates that a subpopulation allele proportion is likely to be close to 0·2, and is almost certainly between 0·1 and 0·3. More precise intervals can be evaluated by computing areas under the beta density curve. This can be done easily in a statistical computer package such as R (available free at www.r-project.org; use function pbeta) or via a numerical approximation based on (5.5).

The beta distribution applies exactly under various theoretical models. It is unlikely to be strictly valid in practice, but it usually provides a good approximation and allows the essential features of genetic differentiation to be modelled and estimated in actual populations.

The beta-binomial sampling formula

If the beta is adopted for the distribution of the allele proportion in a subpopulation, then a convenient recursive formula becomes available for the probabilities of samples drawn at random from the subpopulation. Suppose that n alleles have been sampled in the subpopulation, of which m are green. Then the probability that the next allele sampled is also green is:

$$\frac{m\theta + (1-\theta)p}{1 + (n-1)\theta}. \tag{5.6}$$

When $m = n = 0$, we obtain probability p that the first allele drawn is green. The probability that the first two alleles drawn are both green is

$$p(\theta + (1-\theta)p) = p^2 + \theta p(1-p). \tag{5.7}$$

Increasing θ raises the probability of observing two green alleles, because the first observation of a green allele suggests that they are relatively common in the subpopulation, and so drawing another green allele is less surprising. Expressed another way: once we see one green allele we expect that it will have close relatives nearby, so that seeing another green allele becomes more likely. If there were no subpopulation variation, the second green allele would have the same probability as the first, and so the probability of two green alleles would be p^2, obtained by substituting $\theta = 0$ in (5.7). Increasing θ also increases the probability of two blue alleles but decreases the probability of a green allele followed by a blue, which is:

$$(1-\theta)p(1-p). \tag{5.8}$$

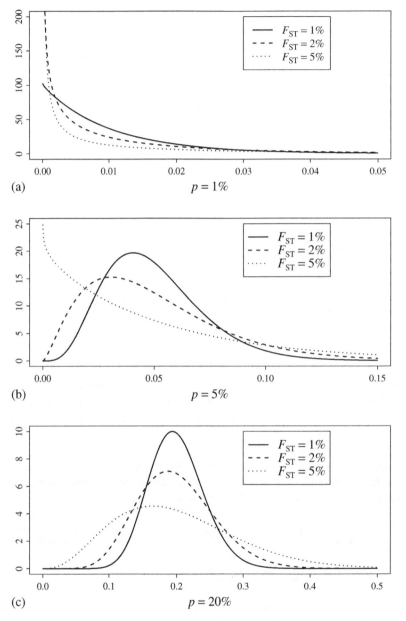

Figure 5.2 Probability density curves for the beta distribution when $p = 0.01$ (a), $p = 0.05$ (b) and $p = 0.20$ (c), and $\lambda = 99$, 49, and 19, so that $\theta\ (= F_{ST}) = 1\%$, 2%, and 5%.

Formula (5.6) can be used recursively to obtain the probability $P(m, n-m)$ of m green alleles in a sample of size n:

$$P(m, n-m) = \frac{\Gamma(\lambda)}{\Gamma(n+\lambda)} \frac{\Gamma(m+\lambda p)}{\Gamma(\lambda p)} \frac{\Gamma(n-m+\lambda(1-p))}{\Gamma(\lambda(1-p))}. \quad (5.9)$$

Here, $\Gamma(x+1) = x\Gamma(x)$, so that (5.9) can be expanded to obtain

$$P(m, n-m) = \frac{\prod_{i=0}^{m-1}(i+\lambda p) \prod_{i=0}^{n-m-1}(i+\lambda(1-p))}{\prod_{i=0}^{n-1}(i+\lambda)}. \quad (5.10)$$

The $\prod$ symbol is similar to $\sum$ but stands for multiplication rather than addition, and so

$$\prod_{i=j}^{k} h(i) = \begin{cases} 0 & \text{if } j > k, \\ h(j) & \text{if } j = k, \\ h(j) \times h(j+1) \times \cdots \times h(k) & \text{if } j < k. \end{cases}$$

On substituting $(1/\theta) - 1$ for λ in (5.10) and simplifying, we obtain

$$P(m, n-m) = \frac{\prod_{i=0}^{m-1}(i\theta + (1-\theta)p) \prod_{i=0}^{n-m-1}(i\theta + (1-\theta)(1-p))}{(1-\theta)\prod_{i=1}^{n-2}(1+i\theta)}. \quad (5.11)$$

Formula (5.11) applies to ordered samples. For unordered samples, it should be multiplied by $\binom{n}{m}$, the number of ways of choosing m objects from n, so that, for example, if $n = 4$ and $m = 2$, then $\binom{n}{m} = 6$. The resulting formula is called the *beta-binomial* sampling formula.

Formula (5.9) is concise and convenient for computation, because there exist fast algorithms for evaluating the Γ function. Although (5.11) may seem daunting at first sight, it is easy to work with after a little practice. For example, the probability of a green allele followed by a blue is the case $m = 1, n = 2$:

$$P(1, 1) = \frac{\prod_{i=0}^{0}(i\theta + (1-\theta)p) \times \prod_{i=0}^{0}(i\theta + (1-\theta)(1-p))}{1-\theta}$$

$$= \frac{(1-\theta)p \times (1-\theta)(1-p)}{1-\theta}$$

$$= (1-\theta)p(1-p),$$

in agreement with (5.8). To obtain the probability of an unordered sample of one green and one blue allele, we need to multiply by $\binom{2}{1} = 2$.

If you remain unpersuaded that (5.11) is not too hard to use, you can forget it: all formulas required in this book can be derived by successive applications of the simpler, recursive formula (5.6).

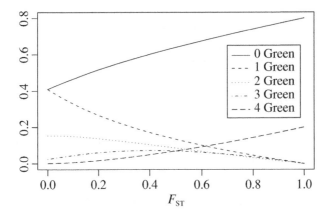

Figure 5.3 Probabilities for the number of "green" alleles in a sample of size four under the beta-binomial sampling formula with $p = 0{\cdot}2$ and θ ($= F_{ST}$) ranging from 0 to 1.

Samples of size four

We will see in Chapter 6 that samples of size four from a subpopulation have a special importance in forensic identification problems. Using (5.11), the probability of two green alleles followed by two blue ones is

$$P(2,2) = \frac{\prod_{i=0}^{1}(i\theta + (1-\theta)p) \times \prod_{i=0}^{1}(i\theta + (1-\theta)(1-p))}{(1-\theta)\prod_{i=1}^{2}(1+i\theta)}$$

After simplifying and multiplying by $\binom{4}{2}$, we obtain the probability of an unordered sample of two green and two blue alleles:

$$6p(1-p)\frac{(1-\theta)(\theta + (1-\theta)p)(\theta + (1-\theta)(1-p))}{(1+\theta)(1+2\theta)}, \quad (5.12)$$

which, for $p = 0{\cdot}2$, is shown as the dotted curve in Figure 5.3. At $\theta = 0$, this curve has height $6p^2(1-p)^2 = 0{\cdot}1536$; it rises slightly to $0{\cdot}1540$ at $\theta = 1\%$, but subsequently starts to decline, reaching $0{\cdot}1539$ at $\theta = 2\%$ and $0{\cdot}1535$ at $\theta = 5\%$, and eventually 0 at $\theta = 1$.

The probability that all four alleles are green is

$$P(4,0) = \frac{\prod_{i=0}^{3}(i\theta + (1-\theta)p)}{(1-\theta)\prod_{i=1}^{2}(1+i\theta)}$$

$$= \frac{p\prod_{i=1}^{3}(i\theta + (1-\theta)p)}{(1+\theta)(1+2\theta)}, \quad (5.13)$$

which, for $p = 0{\cdot}2$, is shown as a dashed curve in Figure 5.3. It is the lowest curve in the figure when $\theta = 0$ (height $(0{\cdot}2)^4 = 0{\cdot}0016$), but it increases markedly to $0{\cdot}0020$ and $0{\cdot}0041$, when $\theta = 1\%$ and 5%, and eventually reaches $0{\cdot}2$ at $\theta = 1$.

The larger the value of θ, the closer the subpopulations are to fixation and so the more likely it is that samples are all the same allele. When $\theta = 1$, fixation has occurred in all subpopulations, and we see from Figure 5.3 that the sample is either all-green (probability 0·2) or no-green (probability 0·8).

5.3.2 Multi-allelic loci

Formula (5.6) still holds if there are more than two alleles segregating at the locus. The multivariate extension of the beta distribution for the subpopulation allele proportions is the Dirichlet, which has pdf

$$f(x_1, x_2, \ldots, x_K) = c \prod_{k=1}^{K} x_k^{\lambda p_k - 1}, \tag{5.14}$$

where $p_1, p_2, \ldots, p_K$ denote the population allele proportions, with $\sum_{k=1}^{K} p_k = 1$, and similarly the x_k are all positive and sum to one. If $K = 2$ then $p_2 = 1 - p_1$ and $x_2 = 1 - x_1$ and (5.5) is recovered.

The sampling formula (5.9) now becomes (ordered case)

$$P(m) = \frac{\Gamma(\lambda)}{\Gamma(n + \lambda)} \prod_{k=1}^{K} \frac{\Gamma(m_k + \lambda p_k)}{\Gamma(\lambda p_k)}, \tag{5.15}$$

where $m = (m_1, m_2, \ldots, m_K)$ denotes the vector of allele counts, so that $n = \sum_{k=1}^{K} m_k$. Analogous to (5.11), formula (5.15) can also be expressed in the form

$$P(m) = \frac{\prod_{k=1}^{K} \prod_{i=0}^{m_k - 1} (i\theta + (1-\theta)p_k)}{(1-\theta) \prod_{i=1}^{n-2}(1 + i\theta)}. \tag{5.16}$$

For unordered samples, (5.15) and (5.16) should each be multiplied by $n!$ and divided by the product of the $m_k!$, where $n! \equiv n \times (n-1) \times (n-2) \times \cdots \times 2$ (and $0! = 1! = 1$). For example, $4! = 4 \times 3 \times 2 = 24$.

Suppose that there are three alleles with population proportions p_1, p_2, and p_3 so that $p_1 + p_2 + p_3 = 1$. Under (5.16), the probability that an unordered sample of size three from the subpopulation consists of one copy of each allele is

$$P(1, 1, 1) = 3! \frac{\prod_{k=1}^{3} \prod_{i=0}^{0}(i\theta + (1-\theta)p_k)}{(1-\theta)\prod_{i=1}^{1}(1 + i\theta)}$$

$$= \frac{6}{(1-\theta)(1+\theta)} \prod_{k=1}^{3} (1-\theta)p_k$$

$$= 6p_1 p_2 p_3 \frac{(1-\theta)^2}{1+\theta}.$$

Similarly,

$$P(2,1,0) = 3p_1 p_2 (1-\theta)\frac{(\theta + (1-\theta)p_1)}{1+\theta},$$

$$P(3,0,0) = p_1(\theta + (1-\theta)p_1)\frac{(2\theta + (1-\theta)p_1)}{1+\theta}. \qquad (5.17)$$

The latter two formulas are the same whether the locus is diallelic or multi-allelic.

The sampling formula (5.16) is very useful in population genetics, and especially for forensic applications. It has been derived under the assumption that subpopulation allele proportions follow the Dirichlet distribution (which includes the beta distribution when $K = 2$). This assumption holds in some simple population-genetics models, and in Section 5.6 below we outline a derivation of (5.16) that does not require the Dirichlet distributional assumption.

Nevertheless, (5.16) cannot be regarded as exact in practice. Marchini *et al.* (2004) found that the beta-binomial sampling formula provided an excellent fit for a genome-wide study of SNP markers (Section 4.5). However, for STR loci, mutation is such that the mutant allele usually differs from its parent by exactly one repeat unit (see Section 5.1.2), and this makes it unlikely that the Dirichlet assumption will be strictly valid if mutation is important relative to migration in explaining geographical patterns of STR allele proportions. More generally, the complex patterns of mating and migration of natural populations make it implausible that any mathematical formula can be regarded as precise. Instead, the question is: how good is the approximation for the purpose at hand? Formula (5.16) captures the most important effects of drift in a subdivided population. See Weir (1996) and Balding (2003) for further discussion.

5.4 Hardy–Weinberg equilibrium

So far, we have been ignoring the fact that, at autosomal loci, genes come in pairs, one maternal in origin, the other paternal. Ignoring this fact is justified under the assumption of Hardy-Weinberg Equilibrium (HWE), which refers to the independence of an individual's two alleles at a locus. If HWE holds in an infinitely large population, then the genotype proportions in the homozygote and heterozygote case are of the form

Genotype:	GG	BG
Hardy-Weinberg proportion:	p_G^2	$2p_B p_G$

If HWE holds, then knowing one of the genotype proportions at a diallelic locus, we can predict the two unknown proportions. For example, if 4% of the population is affected by a fully penetrant recessive Mendelian trait, then we have $P_{GG} = 0.04$ and so $p_G = 0.2$ and $p_B = 0.8$. Under HWE, it follows that 32% of the population ($= 2p_B p_G$) is an unaffected carrier, that is has the BG genotype. Thus, a person unaffected by the trait has probability $32/96 = 1/3$ of being a carrier.

The ability to make such predictions was one of the first successes for population genetics nearly 100 years ago, and remains useful today.

HWE never holds exactly in real populations, but it provides a good approximation if the population size is large, if mating is at random in the population, and if there is no differential survival of zygotes according to their genotype at the locus (i.e. no selection). If, however, the population is subdivided, then drift can cause it to deviate from HWE even if neutrality and random mating hold within subpopulations. Because an individual's two alleles form a sample of size two from a subpopulation, formula (5.16) applies (unordered case) and we obtain the genotype proportions

Genotype:	GG	BG
Proportion:	$p_G^2 + f p_G(1 - p_G)$	$2(1 - f)p_B p_G$

where we have followed the convention of using f ($= F_{IT}$) in place of θ ($= F_{ST}$) when the two genes are drawn from the same individual, although $f = \theta$ when, as we have assumed here, HWE applies within the subpopulations.

Inbreeding refers to a pattern of mate choice such that mates tend to be more closely related than random pairs of individuals within the population. Inbreeding in an unstructured population leads to genotype proportions that are the same as those in a non-inbreeding, structured population given above, where f is now interpreted as the probability of identity by descent of the two genes uniting in a gamete. Indeed, the coancestry interpretation of θ (see Section 5.6 below) clarifies that population structuring and inbreeding have similar effects on individual genotype proportions. We briefly discuss the combined effects of both population structure and inbreeding on DNA profile match probabilities in Section 6.2.2 below.

Assortative mating refers to a practice of mate choice based on phenotype, for example, when mates tend to be more similar (e.g. in height, skin colour, or intelligence) than are random pairs from the population. This affects genotype proportions in the same way as does inbreeding, except that it is limited to loci involved in determining the phenotype.

Other than population structure, inbreeding, and assortative mating, the most important cause of deviations away from HWE is selection. In the extreme case of a lethal recessive allele, the homozygote genotype may be absent from the adult population even though the allele persists in heterozygote form.

In practice, HWE holds approximately in most human populations and at most loci, and often the reason for deviation from HWE in observed samples is genotyping errors rather than deviation from HWE in the underlying population. For example, the failure to observe one allele in a heterozygote (Section 4.1.1) may lead to the genotype being wrongly recorded as a homozygote for the other allele.

5.4.1 Testing for deviations from HWE †

We argue in Section 5.4.2 and 6.2.1 that testing for deviations from HWE is not as important for forensic purposes as is often believed. However, it is nevertheless

SOME POPULATION GENETICS FOR DNA EVIDENCE

Table 5.1 Pearson's test of HWE for a small dataset.

Genotype	GG	BG	BB	total
Observed (O)	10	5	5	20
Expected (E)	7·8	9·4	2·8	20
$(O-E)^2/E$	0·61	2·04	1·70	4·36

worthwhile, at least to check for genotyping or data recording errors, and so we briefly introduce some approaches to testing.

We describe and illustrate Pearson's χ^2 goodness-of-fit test and Fisher's exact test. Pearson's is the easiest test to apply, though Fisher's is usually superior in practice – see the discussion of Maiste and Weir (1995). Often, a better alternative is to estimate a parameter measuring divergence from HWE, and we also briefly introduce this approach below.

Pearson's test

Consider the small dataset of genotypes at a diallelic locus shown in the "observed" row of Table 5.1. The total allele counts are 25 G and 15 B, leading to an estimate of 25/40, or 5/8, for p_G. Using this estimate, the genotype probabilities *if* HWE holds are estimated to be:

$$P(\text{GG}) \approx (5/8)^2 = 25/64$$
$$P(\text{BG}) \approx 2 \times (5/8) \times (3/8) = 30/64$$
$$P(\text{BB}) \approx (3/8)^2 = 9/64.$$

The corresponding expected counts in a sample of size 20 are then approximately 7·8, 9·4, and 2·8. These are not close to the observed counts, but is the discrepancy large enough that we should be convinced that HWE does not hold?

Pearson's goodness-of-fit statistic provides an answer to this question. The statistic can be computed as the sum of the squared differences between observed and expected values, each divided by the expected value (see Table 5.1). Alternatively, in the diallelic case, there is a shortcut formula:

$$n \left(\frac{ac - (b/2)^2}{(a+b/2)(b/2+c)} \right)^2,$$

where a, b, and c denote the counts of the GG, BG and BB genotypes respectively. Here, $a = 10$, $b = c = 5$, and we obtain

$$20 \times \left(\frac{50 - 6 \cdot 25}{12 \cdot 5 \times 7 \cdot 5} \right)^2 = 4 \cdot 36,$$

which is greater than 3·84, the 95% point of the χ_1^2 distribution, but less than 6·63, the 99% point. Thus, at the 5% level of significance, we can conclude that HWE does not hold in the population from which the sample was drawn, but we cannot reject HWE at the 1% significance level.

For a K-allele locus, the number of degrees of freedom of the χ^2 distribution is the number of heterozygote genotypes, which is $K(K-1)/2$. The expected value of a χ_m^2 distribution is m, and so if the test statistic is less than m, we know without looking up tables that the null hypothesis cannot be rejected. The χ^2 distribution for Pearson's statistic is only approximate, and the approximation is poor if some of the expected genotype counts are small. As a rule of thumb, most of the expected counts should exceed five and all should exceed two.

Pearson's test can be implemented using the chisq.test function of R.

Fisher's exact test

Fisher's exact test of deviation from HWE is typically more powerful than the Pearson test and does not rely on the χ^2 approximation.

Consider a diallelic locus, at which the genotype counts observed in a sample of size n are n_{GG}, n_{BG}, and n_{BB}. Write n_G and n_B for the allele counts and p_G and p_B for population allele proportions. Under HWE, the probability of the genotype counts *given* the allele counts is

$$P(n_{GG}, n_{BG}, n_{BB} \mid n_G, n_B) = \frac{n! n_G! n_B! 2^{n_{BG}}}{(2n)! n_{GG}! n_{BG}! n_{BB}!} \quad (5.18)$$

(Remember that $n! \equiv n \times (n-1) \times (n-2) \times \cdots \times 2$, and $0! = 1! = 1$).

Fisher's exact test uses the probability (5.18) as the test statistic. Consider all the possible ways of reassigning the observed alleles into genotypes. For example if we observed $n = 2$ individuals with genotypes GG and BB, then there are just two possible genotype assignments with probabilities:

Genotypes	Probability
GG, BB	$(2!2!2!2^0)/(4!1!0!1!) = 1/3$
BG, BG	$(2!2!2!2^2)/(4!0!2!0!) = 2/3$

The p-value of the test is the total probability of all genotype assignments that are as or less probable, according to (5.18), than the observed assignment.

For the example introduced in Table 5.1, there are just eight possible genotype assignments given 25 G and 15 B alleles, and these are shown in Table 5.2. The genotype assignments with 1, 3, and 15 heterozygotes are each less likely than the observed assignment, and they have total probability 0·016. In small testing problems, it is customary to halve the probability assigned to the observed value – this is not necessary in practice because the probability of any particular observation is usually very small. Here, it leads to a p-value of just under 3·5%, similar to that of Pearson's test.

Table 5.2 Calculations for Fisher's exact test for the dataset of Table 5.1.

n_{GG}	n_{BG}	n_{BB}	Probability	Contribution to p-value
12	1	7	0·0004	0·0004
11	3	6	0·0028	0·0028
10	5	5	0·0370	0·0185
9	7	4	0·1764	
8	9	3	0·3527	
7	11	2	0·3078	
6	13	1	0·1105	
5	15	0	0·0126	0·0126
			1·0000	0·0343

In practice, evaluating all genotypes consistent with the allele counts often is not feasible, and the p-value is instead approximated via Monte Carlo methods. The test is then no longer exact, but the Monte Carlo approximation can be made arbitrarily close with sufficient iterations. Guo and Thompson (1992) describe two such methods, the first based on random pairings of the observed alleles while the second is a Markov chain Monte Carlo method which, by making only small changes in the genotype assignments, avoids full computation of (5.18) at each step.

Fisher's test can be implemented using the fisher.test function of R.

Estimating the inbreeding parameter

To keep the discussion simple, here we will restrict attention to a diallelic locus and assume $p_G = 1 - p_B$ to be known. See Ayres and Balding (1998) for Bayesian estimation of f allowing for uncertainty in the population allele proportions, and for multi-allelic extensions.

Under the inbreeding model, the homozygote and heterozygote probabilities are

$$P(GG) = p_G^2 + f p_B p_G$$
$$P(BG) = 2(1-f) p_B p_G$$
$$P(BB) = p_B^2 + f p_B p_G,$$

where $\max(-p_B/p_G, -p_G/p_B) \le f \le 1$. Under this model, the likelihood of a sample with genotype counts n_{GG}, n_{BG}, and n_{BB} is

$$L(f) = cP(GG)^{n_{GG}} P(BG)^{n_{BG}} P(BB)^{n_{BB}},$$

where c is an arbitrary constant. One possibility is to choose c such that $L(f)$ takes value one at the HWE value $f = 0$, in which case we obtain

$$L(f) = (1 + f p_B/p_G)^{n_{GG}} (1-f)^{n_{BG}} (1 + f p_G/p_B)^{n_{BB}}.$$

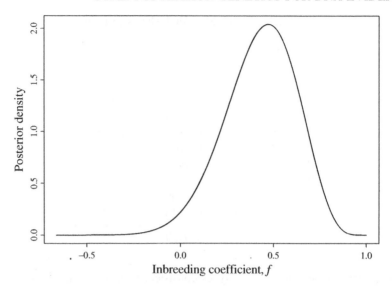

Figure 5.4 Posterior probability density for the inbreeding coefficient f given sample genotype counts $n_{GG} = 10$, $n_{BG} = 5$, and $n_{BB} = 5$, $p_G = 0.6$, $p_B = 0.4$, and a uniform prior.

If instead c is chosen so that the integral over f is one, then the likelihood also specifies the posterior density curve for f given a uniform prior distribution. Figure 5.4 plots this density curve for the data of Table 5.1, assuming p_G and p_B equal to their sample proportions.

Bayesian hypothesis testing is usually performed via the Bayes factor, which is the probability of the observed data under the inbreeding model divided by its probability under HWE. Here, the Bayes factor turns out to be about 2.7 in favour of the inbreeding model, but this depends sensitively on the choice of a uniform prior distribution for f. An informal Bayesian hypothesis test, less sensitive to the prior, can be achieved by looking at whether the HWE value falls within a given interval of highest posterior density for f. Here, the 95% highest posterior density interval is (0.050, 0.784), excluding $f = 0$, but the 99% interval is (−0.085, 0.849) which does include the HWE value. The results of this test are thus broadly in line with those of the Pearson's and Fisher's tests above, but in addition, we have Figure 5.4, which gives a visual representation of the values of f supported by the data. Because of the small sample size, these values span a wide interval, but the most likely value (the maximum-likelihood estimator) is $\widehat{f} = 0.47$, suggesting a large deviation from HWE.

5.4.2 Interpretation of test results

Care must be taken with the interpretation of the results of any test, since the magnitude of any deviation of HWE is confounded with sample size: a false

SOME POPULATION GENETICS FOR DNA EVIDENCE   75

null hypothesis of HWE may well be accepted when the sample size is small. Conversely, HWE is never strictly true, and given a large enough sample size, it will be rejected even when the magnitude of the deviation from equilibrium is too small to be of any practical significance. An additional potential pitfall in hypothesis testing relates to the problem of multiple testing. If many loci are tested, by chance some will indicate significant disequilibrium even if all loci are in fact in HWE.

It is widely believed that HWE in the population from which a population database is drawn is required for the validity of DNA profile match probability calculations using that database. This is incorrect. Population subdivision and inbreeding both increase the probability of shared ancestry (or "coancestry") of the two genes at a locus and are one of the main potential causes of deviation from HWE. However, their effect on forensic match probabilities can be accounted for using θ, as discussed below in Chapter 6. Deviation from HWE due to assortative mating and/or selection is expected to be small at forensic loci; even if there is some selection effect due to linkage disequilibrium with a functional locus or regulatory sequence, it is likely to be limited to one or two loci of the DNA profile.

Consequently, testing for HWE is not crucial for forensic work. A HWE test is easy to perform, and can signal genotyping or data entry errors, and so is probably worthwhile as a routine check. For forensic applications, accepting the null hypothesis of HWE is no reason for complacency, because we are concerned with the joint probabilities of four alleles, not just two. Conversely, rejecting the null is not necessarily a cause for concern, since this may represent non-HWE due to population subdivision, which is accounted for in the likelihood ratio via θ.

5.5 Linkage equilibrium

Linkage equilibrium (or gametic equilibrium) is the term used in population genetics for the independence of the alleles at different loci in the same gamete[2] at distinct loci. There are many possible causes of linkage disequilibrium (LD). For non-forensic applications, the most important cause is linkage: the tendency for alleles at loci close together on a chromosome to be passed on together over many generations because recombinations between them rarely occur. In a simple, deterministic population-genetics model, the LD between two loci decreases exponentially over time at a rate proportional to the recombination fraction. Although often cited in textbooks, in practice, there is typically so much stochastic noise that the exponential pattern of decay is hardly perceptible. The LD in actual human populations usually extends over a few tens of Kb. Occasionally, it extends over several hundred Kb, but even this is very small compared to a typical chromosome length of around a hundred thousand Kb. Marker loci used in forensic work are chosen sufficiently far apart so that LD due to linkage is negligible for unrelated individuals, except for multiple markers from the Y or the mitochondrial

---

[2]That is, inherited from the same parent.

chromosomes, for which recombination is (almost) absent. Linkage can be important when the DNA evidence includes the profiles of several related individuals (Section 7.1.4).

Other than linkage, LD can be caused by population subdivision and drift. If a population is subdivided, drift may cause the allele proportions at many loci to differ from the population proportions, leading to statistical associations between unlinked loci. Another possible cause is admixture, in which a hybrid population is formed by migrants from several ancestral populations. For both subdivision and admixture, an individual with an allele at one locus that is frequent in one of the subpopulations (or source populations) may have a high level of ancestry from that (sub)population, and hence at a second locus s/he may also have an allele common in that (sub)population.

In forensic match probabilities, LD due to subdivision and drift is accounted for via θ. LD due to admixture is slightly different, in that an admixed population need not have any subdivision, and so the definition of θ in terms of subpopulation allele proportions does not apply. However, whereas subdivision and drift can sustain LD between unlinked loci, this is not true for a random-mating hybrid population, in which LD decays markedly in a small number of generations. Moreover, if a defendant has coancestry with some of the alternative possible culprits in a migrant gene pool, the interpretation of θ as the probability that two alleles are ibd (Section 5.6) clarifies that it can also be used to adjust for this possibility.

Selection is another possible cause for LD at unlinked loci. Whatever the cause of LD, it only matters for forensic applications if it tends to enhance match probabilities. If selection is environment specific, and hence tends to vary with geography, its effect is similar to that of drift and can be accounted for using θ. Other forms of selection are not expected to systematically affect forensic match probabilities.

The principles for, and caveats associated with, testing for deviations from HWE also apply to testing for LD. The tests available differ according to whether genotype or haplotype data are available. Multi-locus genotype data consists of two alleles at each locus. A haplotype includes one allele from each locus, these alleles having the same gamete (i.e. parent) of origin. Each genotype thus corresponds to two haplotypes.

Consider first haplotype data at two diallelic loci, with alleles B and b, and G and g, respectively. Pearson's statistic for testing the hypothesis of linkage equilibrium is nr^2, where r denotes the sample correlation coefficient defined by

$$r = \frac{ad - bc}{\sqrt{(a+b)(c+d)(a+c)(b+d)}},$$

and a, b, c, and d denote the sample counts of BG, Bg, bG, and bg haplotypes, respectively. Sample sizes are usually sufficient to test for independence at only two or three loci using Pearson's test. See Zaykin *et al.* (1995) for a discussion of exact tests.

For testing deviations from LD using genotype data, see Weir (1996) and Schaid (2004).

5.6 Coancestry †

In this section, we derive the sampling formula (5.16) in a simple population-genetics model without explicitly invoking the beta or Dirichlet distributions. It follows that, although θ was introduced above in terms of the variation in subpopulation allele proportions, it can also be interpreted as the probability that two alleles are descended identically from a common ancestor. Hence θ is also called a "coancestry coefficient" or "kinship coefficient".

We adopt a simple model in which the population size is a large, constant N (alleles), and in each generation, each allele has probability u/N that it has arisen as a novel selection from a gene pool such that it is "green" with probability p. Otherwise, it is an exact copy of one of the N alleles in the previous generation, each equally likely to be the parent. "Selection from the gene pool" can be thought of as representing either mutation or migration, or both. Under this model, we outline a derivation of particular cases of the sampling formula for samples of size up to three, and then give a general recursive argument.

One allele: In tracing the ancestry of an allele backwards in time, eventually its ancestor will have been selected from the gene pool, and hence the probability that it is green is $P(1, 0) = p$.

Two alleles: Tracing the ancestry of two alleles backwards in time, in each generation there is probability $1/N$ that they have a common ancestor, and probability $2u/N$ that one of them is selected from the gene pool. Thus, the probability that the two alleles are ibd from a common ancestor in any generation is $1/(1 + 2u)$, and the probability that this ancestor was green is again p. Reasoning similarly for the probability that the two alleles were both drawn at random from the gene pool and are both green, and summing these two terms, gives

$$P(2, 0) = \frac{p + 2up^2}{1 + 2u}$$

for the overall probability that both alleles are green which, on replacing $1/(1 + 2u)$ with θ, is the same as (5.7).

Three alleles: The probability that two of the alleles meet in a common ancestor without any of them having been drawn from the gene pool is

$$\frac{3/N}{3/N + 3u/N} = \frac{1}{1+u} = \frac{2\theta}{1+\theta}.$$

Continuing back in time, the probability that the remaining two alleles also meet in a common ancestor is θ, and so the probability that all three alleles are ibd is

$$\frac{2\theta^2}{1+\theta}.$$

Similarly, it can be seen that the probabilities of exactly one and zero pairs of alleles being ibd from a common ancestor in the subpopulation are

$$\frac{3\theta(1-\theta)}{1+\theta} \quad \text{and} \quad \frac{(1-\theta)^2}{1+\theta}.$$

The probability that three alleles sampled are all green is then

$$P(3,0,0) = p\frac{2\theta^2}{1+\theta} + p^2\frac{3\theta(1-\theta)}{1+\theta} + p^3\frac{(1-\theta)^2}{1+\theta}$$

which simplifies to give (5.17).

General proof via recursion: For ease of presentation, we restrict attention to the unordered form of the diallelic sampling formula (5.11), but a similar argument applies for the multi-allelic formula (5.16). Consider tracing back the ancestry of a sample of n alleles, of which m_1 are green and m_2 $(= n - m_1)$ are not. With probability $(n-1)/(n-1+2F)$, the most recent event was a pair of alleles coalescing in a common ancestor; without any allele having been drawn from the gene pool. In this case, the current sample has m_1 green alleles if either

- the ancestral allele was green, and was one of $m_1 - 1$ green alleles in the ancestral sample, or

- the ancestral allele was non-green and the ancestral sample had m_1 green alleles.

Similarly, in the case that the most recent event was a draw from the gene pool, in order that the current sample have m_1 green alleles, either

- the new draw was green and there were previously $m_1 - 1$ green alleles, or

- the newly drawn allele was non-green, and there were previously m_1 green alleles.

Summing these four terms and substituting $\theta = 1/(1+2F)$, we obtain the recursive equation:

$$\begin{aligned}
P(m_1, m_2) &= \frac{\theta}{1+(n-2)\theta}[(m_1-1)P(m_1-1, m_2) + (m_2-1)P(m_1, m_2-1)] \\
&\quad + \frac{1-\theta}{1+(n-2)\theta}[pP(m_1-1, m_2) + (1-p)P(m_1, m_2-1)] \\
&= \frac{(m_1-1)\theta + (1-\theta)p}{1+(n-2)\theta}P(m_1-1, m_2) \\
&\quad + \frac{(m_2-1)\theta + (1-\theta)(1-p)}{1+(n-2)\theta}P(m_1, m_2-1). \quad (5.19)
\end{aligned}$$

It is readily verified that the unordered form of (5.11) satisfies (5.19), provided that $P(i, j)$ is interpreted as zero if either $i < 0$ or $j < 0$.

5.7 Likelihood-based estimation of θ †

We briefly discussed a simple method-of-moments estimator of θ in Section 5.2. In this book, the sampling formula (5.16) will mainly be used to calculate likelihood ratios in Chapter 6, but we note here that it can also be used for likelihood-based estimation of θ.

First, to keep matters simple, we restrict attention to a single, diallelic locus and assume that the reference allele proportions are known. Specifically, we consider a sample of 10 alleles of which 6 are green and suppose that $p_G = 0.2$. The sample proportion 6/10 is much larger than the expected proportion p_G, suggesting that θ is rather large, but any inferences must be very weak with so little data. To make this precise, we use (5.16), or equivalently for a diallelic locus we use (5.11) to obtain:

$$L(\theta) = \frac{\prod_{i=0}^{5}(i\theta + (1-\theta)/5) \times \prod_{i=0}^{3}(i\theta + 4(1-\theta)/5)}{(1-\theta)\prod_{i=1}^{8}(1+i\theta)}$$

This curve is plotted in Figure 5.5 (solid line). As expected, a wide range of θ values are supported. The curve has been scaled so that it can be interpreted as a posterior density given a uniform prior for θ, and the 95% highest posterior density interval for θ runs from 0.027 to 0.83.

Figure 5.5 also shows the corresponding curve when the sample size is increased by a factor of 10, with the sample proportion of green alleles unchanged. Inferences

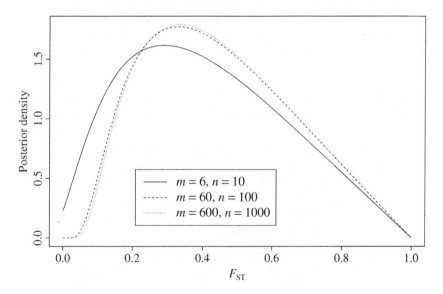

Figure 5.5 Likelihood curves for θ (= F_{ST}) from a sample in one subpopulation, at a diallelic locus with $p = 0.2$. The curves have been scaled so that they can also be interpreted as posterior density curves given a uniform prior for θ.

are now a little stronger: the 95% highest posterior density interval is $0.099 \leq \theta \leq 0.84$, excluding a larger interval near zero.

Stepping up the sample size by a further factor of 10 (dotted curve), the posterior density curve is almost unchanged. This reflects the fact that it is only the subpopulation allele proportion that is informative about θ, and once the sample size is large enough for this to be estimated well, there is no additional benefit from increasing the sample size for that subpopulation and locus.

Instead of increasing the sample size at a particular locus/subpopulation, better inferences about θ can be obtained by combining information across loci and/or across subpopulations, but without assuming that θ is constant. FSTMET is a C program for Bayesian estimation of θ from samples of STR alleles from several loci and several subpopulations that can be obtained from

www.rdg.ac.uk/statistics/genetics

See Balding (2003) for the theory, and Balding *et al.* (1996), Balding and Nichols (1997), and Ayres *et al.* (2002) for some applications to forensic STR data.

I used FSTMET to analyse the data of Butler *et al.* (2003) at 15 forensic STR loci (the 13 CODIS core loci plus D19S433 and D2S1338) from three populations: US Hispanic ($n = 140$), African-American ($n = 256$) and US Caucasian ($n = 301$). This analysis is for illustration only and is not directly applicable to forensic inference. This is because separate databases are maintained for estimating allele proportions in these populations. Although there are issues concerning the representativeness of the databases, the key question of practical importance concerns subpopulation stratification within these major populations. Any such stratification is expected to be less marked than between the major population groups considered here.

First (Table 5.3, row (i)), the samples were treated symmetrically, and the reference allele proportions were regarded as unknown and automatically estimated by FSTMET. Next (Table 5.3, rows (ii)–(iv)), the sample proportions for each

Table 5.3 Posterior mean θ values at 15 forensic STR loci for US Hispanic, African-American and US Caucasian populations using FSTMET applied to the data of Butler *et al.* (2003). The reference allele proportions are (i) estimated from the data, or equal to the (ii) US Caucasian, (iii) African-American, and (iv) US Hispanic, sample proportions.

Population:	US Hispanic	African-American	US Caucasian
(i)	0.009	0.020	0.018
(ii)	0.007	0.026	–
(iii)	0.027	–	0.035
(iv)	–	0.025	0.013

SOME POPULATION GENETICS FOR DNA EVIDENCE   81

population in turn were treated as the reference values, with θ values for the other two populations being estimated relative to these. Here we note that choice of reference values can have an important effect on θ estimates (posterior mean values are shown in the table). Further implications of these analyses are discussed below in Section 6.3.2.

5.8 Population genetics exercises

Solutions start on page 164.

1. The population proportion of an allele is $p = 0.15$. The population is subdivided and $\theta = 2\%$ for each subpopulation.

    (a) What is the standard deviation of the allele proportion in the subpopulations?

    (b) Find an approximate 95% interval for the subpopulation allele proportion (if possible, use the beta distribution, otherwise use a simple approximation).

2. You observe in large samples from five subpopulations the following proportions of an allele at a diallelic locus: 0·11, 0·15, 0·08, 0·09, 0·12.

    (a) You know that $p = 0.1$. Estimate the value of θ (assumed the same for each subpopulation).

    (b) How would your answer from (a) be affected if p was unknown?

3. Use the sampling formula (5.11) to evaluate the probabilities of

    (a) $m = 3$ and   (b) $m = 2$

    G alleles in an unordered sample of size $n = 3$ (alleles) from a haploid population at a diallelic locus. Express your answers in terms of p (the population proportion of G) and θ, and then evaluate them when $p = 0.75$ and $\theta = 0$, 2%, and 10%. (Hint: as a check, verify that you obtain 0·25 and 0 when $\theta = 1$.)

4. Test for deviation from HWE in the population from which the following genotype proportions were observed at a triallelic locus:

Homozygotes:	BB 5	GG 10	RR 5
Heterozygotes:	BG 15	BR 10	GR 20

5. If the observed haplotype frequencies at two diallelic loci are

Haplotype:	BG	Bg	bG	bg
Frequency:	10	20	5	8

    does linkage equilibrium hold in the population at these loci?

6

Identification

6.1 Choosing the hypotheses

We saw in Section 3.2 that the weight-of-evidence formula requires the likelihood ratio R_i for different individuals i other than the defendant. A likelihood ratio (introduced in Section 3.1) measures the weight of evidence in favour of one hypothesis relative to a rival hypothesis. It is not an absolute measure of evidential weight, but depends on how you formulate the competing hypotheses: there are often a number of different hypothesis pairs that might be considered, each with its own likelihood ratio.

In a typical criminal case, the hypotheses of direct interest to the court concern the guilt or innocence of the defendant, s. In some cases, it may not be clear that a crime has occurred, and if so, whether it had a single perpetrator, but when these conditions do hold, the hypotheses of interest are

G : s is the culprit;

I : s is not the culprit.

The hypothesis pair (G, I) is clear and concise, but we noted above in Section 3.5.1 that it is not a practical choice for quantitative evaluation of evidence: no likelihood ratio can be directly computed without a further assumption allowing probabilities to be specified under hypothesis I. Instead, in Section 3.5.1 we discussed the partition of I into a union of events of the form $C = i$, where C is the culprit and i denotes an individual other than s.

Replacing I with $\cup_i \{C = i\}$ is often convenient, but it is not logically necessary. In complex settings, such as those involving multiple contributors to the crime sample (see Section 6.5), the selection of hypotheses may be far from straightforward. In choosing hypotheses to consider, there is almost always a tension between

---

Weight-of-evidence for Forensic DNA Profiles David Balding
© 2005 John Wiley & Sons, Ltd ISBN: 0-470-86764-7

IDENTIFICATION

allowing for all the realistic alternative possible explanations and trying to keep the formulation simple enough to be practical.

There can also be a tension between addressing the hypotheses most relevant to the court's task and respecting the boundary of the forensic scientist's domain of expertise. Commenting directly on the hypothesis pair (G, I) would under some legal systems expose a forensic scientist to the criticism that s/he has offended the "ultimate issue rule". This "rule" seems not to have a precise statement in law but in the context of a criminal trial it is generally interpreted as prohibiting an expert witness from giving a direct opinion about the guilt or innocence of s. The concern motivating the rule seems well grounded: that the forensic scientist should not usurp the function of a judge or jury nor act as an advocate for the defendant. It can be difficult to demarcate the boundary between acceptable and unacceptable statements, and few legal authorities have attempted to do this (Robertson and Vignaux 1995). Therefore, an overly strict interpretation of the rule could diminish a forensic scientist's ability to give full assistance to jurors in their task of reaching a reasoned opinion on the ultimate issue (see Section 9.3.4). However, a reasonable interpretation of the rule accords with the principles that I have been advocating in this book, and can be helpful. For a further discussion, see Robertson and Vignaux (1995).

A forensic scientist may report likelihood ratios comparing hypotheses such as

G' : s is the source of the crime-scene DNA sample;

I' : s is not the source.

In an allegation of rape, consent may be an issue. This would imply an important difference between G and G', yet the issue of consent would usually lie outside the forensic scientist's domain. Alternatively, the evidence may concern a stain found on the defendant's clothing that could have come from the victim, so that the relevant hypotheses might be

G'' : the victim is the source of the DNA sample;

I'' : the victim is not the source.

A victim's blood could be on the clothing of s even though s/he did not cause the injury, in which case G and G'' are distinct.

For a further discussion of the ultimate issue rule, see Robertson and Vignaux (1995). Cook *et al.* (1998) and Evett *et al.* (2000) introduce hierarchies of hypotheses and discuss a rationale for choosing the appropriate level of the hierarchy. To keep the discussion and notation as simple as we can, where no confusion seems possible we will continue to use (G, I) to denote the hypothesis pair of interest, even though (G', I') is more likely to be the actual hypothesis pair to be addressed by the forensic scientist. The issues are broadly the same if the actual hypothesis pair is (G'', I''), except that in the discussions of relatedness below we would be concerned with the relatedness to the victim of the alternative possible sources of DNA, rather than their relatedness to s.

6.1.1 Post-data equivalence of hypotheses

Meester and Sjerps (2003, 2004) discuss the "two-stain" problem in which a defendant's DNA profile matches that obtained from two distinct crime stains and different likelihood ratios can be obtained by contrasting different pairs of hypotheses. This causes no difficulty provided that the weight-of-evidence formula (3.3) is used to combine the likelihood ratios.

More generally, Dawid (2001, 2004) notes that, once the data have been observed, there can be sets of hypotheses pairs that were distinct *a priori* but which have been rendered equivalent by the data. The hypotheses pairs in such a set are said to be "conditionally equivalent". The pairs may have different likelihood ratios but, because they are conditionally equivalent, the allocation of posterior probabilities must, logically, be the same whichever pair of hypotheses is used for the analysis.

For example, let us return briefly to the island of Section 2.2.1, where life and crime are simple thanks to a number of assumptions. Suppose that two suspects s and s' are investigated, and s is found to have Υ while s' does not. Consider the hypothesis pairs:

$$H_1 : s \text{ is guilty};$$
$$H_2 : s \text{ is innocent};$$

and

$$H_1' : \text{either } s \text{ or } s' \text{ is guilty};$$
$$H_2' : s \text{ and } s' \text{ are both innocent.}$$

These pairs become logically equivalent following the observation of the Υ states of s and s', given the fact that the culprit is a Υ-bearer. Under the assumptions of the island problem, the likelihood ratio for H_1 versus H_2 is double that for H_1' versus H_2'. The prior probability of H_1 is half that of H_1' and, since these two effects cancel, the posterior probability of H_1 equals that of H_1'.

In this example, we obtain the same probability that s is the culprit, whichever pair of hypotheses is contrasted, even though the likelihood ratios differ by a factor of two. Strictly speaking, either pair of hypotheses could be employed by a rational juror to arrive at the same conclusion. Nevertheless, although not a logical necessity, if s is on trial, then the argument that (H_1, H_2) is the most appropriate hypothesis formulation seems compelling, as these hypotheses directly address the question of interest to the court. In contrast, the hypothesis pair (H_1', H_2'), which is conditionally equivalent to (H_1, H_2), is more complex for no obvious gain–it would invite confusion to introduce a likelihood ratio for this hypothesis pair to measure the weight of evidence against s.

Similarly, in simple identification scenarios with a single crime stain assumed to have originated from a single contributor, it is very natural to compare G with all the hypotheses of the form "i is the culprit", for different i. Then there is not

IDENTIFICATION

one likelihood ratio, but several, according to the differing levels of relatedness of i with s. A collection of likelihood ratios for various levels of relatedness might make a useful summary of evidential weight. A single likelihood ratio comparing s with an unrelated i can make a useful summary when close relatives are unequivocally excluded, provided that the complexity of the underlying situation is appreciated.

When s has been identified through having the only matching profile in an intelligence database of DNA profiles (Section 3.4.5), Stockmarr (1999) advocated contrasting the hypotheses

G^* : one of the persons whose profiles are in the database is the culprit;

I^* : not G^*.

Assuming no error, the pair (G^*, I^*) is equivalent to (G, I) given the data that s gives the only match in the database. Therefore, it is possible to work coherently with these hypotheses and arrive at the appropriate probability of guilt. However, because (G^*, I^*) does not address the question of interest to the court, likelihood ratios related to these hypotheses can be seriously misleading if proposed as measures of evidential weight. For example, the likelihood ratio based on (G^*, I^*) depends on database size, which can lead to an erroneous conclusion that the evidential strength of a DNA profile match weakens with increasing database size (see Section 3.4.5). For further discussion and criticisms of Stockmarr (1999), see Donnelly and Friedman (1999), Balding (2002), and references therein.

6.2 Calculating likelihood ratios

6.2.1 The match probability

Assume that the evidence under consideration, E, consists only of the information that the culprit C and the defendant s have the same DNA profile D. The possible effects of other profiled individuals are discussed in Section 6.2.6. Introducing the notation $x \equiv D$ to denote "x has profile D", E can be written succinctly as $C \equiv s \equiv D$, and the likelihood ratio (3.1) becomes

$$R_i = \frac{P(C \equiv s \equiv D \mid C = i, E_o)}{P(C \equiv s \equiv D \mid C = s, E_o)}, \quad (6.1)$$

where E_o denotes the other evidence and background information (we are continuing to assume that the DNA evidence is assessed last).

Here, we initially ignore the possibility of error (until Section 6.3.3). Thus, if $C = s$, then $C \equiv s \equiv D$ is equivalent to just $s \equiv D$. Assuming also that the fact that an individual committed the crime does not of itself alter the probability that they have a particular profile, (6.1) can be simplified further to

$$R_i = \frac{P(i \equiv s \equiv D \mid E_o)}{P(s \equiv D \mid E_o)}$$
$$= P(i \equiv D \mid s \equiv D, E_o). \quad (6.2)$$

Thus, the likelihood ratio R_i reduces to the conditional probability, called the "match probability", that i has the profile *given* that s has it: population-genetic effects arise, and can be dealt with, via this conditioning. Other evidence E_o such as eyewitness reports and alibis are typically irrelevant to DNA profile match probabilities. However, E_o could include

- information about the relatedness of s with some other individuals,
- allele proportion information from population databases of DNA profile, and
- other relevant population-genetics data and theory;

and these are potentially important for match probabilities.

The important feature of the match probability is that it takes account of *both* the observed profiles that form the match. Some authors, including the authors of NRC report (Section 9.4) misleadingly refer to the population proportion of the profile as a "match probability", which is inappropriate since the concept of "match" involves two profiles, rather than just one. Equation (6.2) indicates that the question relevant to forensic identification is not

> "what is the probability of observing a particular profile?"

but

> "given that I have observed this profile, what is the probability that another (unprofiled) individual will also have it?".

To emphasise the distinction, Weir (1994) employs the term "conditional genotype frequency".

In this section, we answer this question using the statistical model and sampling formula introduced in Section 5.3. Recall that the sampling formula follows from the assumption that subpopulation allele proportions have the beta (or Dirichlet) distribution with mean p and variance $\theta p(1-p)$. The value of θ can be interpreted either in terms of the mean square error of subpopulation allele proportions (Section 5.3), or in terms of coancestry in a simple migration-drift or mutation-drift model (Section 5.6).

The parameter θ cannot encapsulate all important population-genetics phenomena, but it does capture the essentials relevant to forensic match probabilities. Selection acting jointly at two or more loci each linked with an STR locus can distort population profile proportions away from estimates based on assuming independence across loci. However, this is *a priori* unlikely, and there is no reason to expect any such effects to systematically favour or disfavour defendants. The only population-genetics phenomenon that, if ignored, systematically disfavours defendants is coancestry between defendant and alternative possible culprits, and that is what θ is intended to account for. Human population genetics is complicated, and inevitably θ is an imperfect measure, but by choosing a suitable value, defendants will not be systematically disfavoured, while match probabilities remain small enough to form the basis of satisfactory prosecutions in most cases.

6.2.2 One locus

Consider the case that both i and s are homozygous for allele A. If we assume that they are unrelated, both come from the same subpopulation, and neither is inbred, then (6.2) corresponds to the conditional probability that two further alleles are both A, given a sample of two alleles that are both A. The recursive form of the sampling formula (5.6) allows us to compute such conditional probabilities directly. Evaluating (5.6) twice, first with $m = n = 2$, then with $m = n = 3$, and multiplying the results, gives

---

Single locus: homozygous case

$$R_i = \frac{(2\theta + (1-\theta)p_A)(3\theta + (1-\theta)p_A)}{(1+\theta)(1+2\theta)} \qquad (6.3)$$

In the heterozygous case, under the same assumptions, we need the probability that two further alleles are A and B, given that two observed alleles are A and B. Evaluating (5.6) with $m = 1, n = 2$ and again with $m = 1, n = 3$, and multiplying by two for the two possible orderings of the A and B alleles, gives:

Single locus: heterozygous case

$$R_i = 2\frac{(\theta + (1-\theta)p_A)(\theta + (1-\theta)p_B)}{(1+\theta)(1+2\theta)} \qquad (6.4)$$

These match probability formulas are conditional on the values of θ and the p_j. Thus, strictly, they only apply if these parameters are known exactly, which is never the case in practice. In a fully Bayesian approach, (6.3) and (6.4) should be integrated (that is, averaged) with respect to probability distributions for the unknown values of θ and the p_j. These probability distributions should be based on the available background information about the parameters, for example, from forensic DNA profile databases as well as population-genetics theory and data. I prefer a simpler and more interpretable approach in which this background information is used to obtain estimates for θ and the p_j, which are then "plugged in" to the conditional formulas. Care is required to choose the most appropriate "plug-in" values for the parameters, as these need not satisfy the usual criteria for statistical estimators (see Section 6.3).

Use of (6.3) and (6.4) implies an assumption of HWE within subpopulations, but not in the broader population from which the forensic database is drawn. This is the reason, discussed above in Section 5.4, that deviations from HWE in the population are not directly an issue in match probability calculations. Such deviations do directly affect the probability of observing a particular genotype, but not the conditional match probability (6.2). Thus, as we have already noted in Section 5.4.1, testing for deviation from HWE within the population is of limited value, whereas testing within the subpopulation is not usually feasible.

Ayres and Overall (1999) have developed match probability formulas that take into account inbreeding within subpopulations. These formulas are more complicated than (6.3) and (6.4), but nevertheless relatively easy to apply. In the presence of within-subpopulation inbreeding, the formulas of Ayres and Overall (1999) indicate that match probabilities are slightly increased for homozygotes and decreased for heterozygotes; the overall effect on profile match probabilities is usually very small, but may be worth taking into account when highly inbred populations are relevant to a case.

Some single-locus match probabilities based on (6.3) and (6.4) are shown in Figure 6.1 for θ ranging from 0 to 1. Notice that increasing θ does not necessarily increase the match probability in the heterozygous case, although this almost always occurs in practice. Whatever the values of p_1 and p_2, as θ approaches 1, the homozygous and heterozygous match probabilities approach 1 and 1/3 respectively. Table 6.1 gives match probabilities for a possible genotype at each of four STR loci and for several values of θ. Very roughly, the single-locus match probabilities with $\theta = 5\%$ are about 50% higher than when $\theta = 0$.

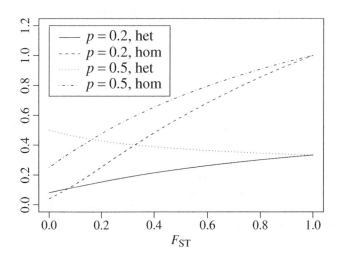

Figure 6.1 Single-locus match probabilities calculated using (6.3) and (6.4) for $\theta (= F_{ST})$ ranging from 0 to 1. In the heterozygote case, the proportions of the two alleles are both equal to p.

IDENTIFICATION

Table 6.1 Single-locus match probabilities assuming i unrelated to s, for four STR loci and various values of θ, for the alleles specified in column 2 and assuming the population proportions given in column 3.

STR locus	Geno-type	Population proportions	Match probability ($\times 10^3$)			
			$\theta = 0$	$\theta = 1\%$	$\theta = 2\%$	$\theta = 5\%$
D18	14, 16	0·16, 0·14	45	49	52	64
D21	28, 31	0·23, 0·07	32	37	41	54
THO1	9·3, 9·3	0·30, 0·30	90	101	112	145
D8	10, 13	0·09, 0·33	59	65	70	85

6.2.3 Multiple loci: the "product rule"

There has been much debate in the forensic science literature about the validity of the "product rule" for combining DNA match probabilities across loci. Combining probabilities via multiplication implies an assumption of statistical independence, and so the debate is equivalent to a debate about the *independence* of STR genotypes at different loci. This question can be rephrased:

> For two distinct individuals i and s does the event that their genotypes match at one or more STR loci affect the probability that they will match at the next locus?

It is important to appreciate that "independence" is not an absolute concept, but can depend on the modelling assumptions employed in deriving the probabilities. In children of mixed ages, reading ability and shoe size are not independent, because older children have, on average, both larger shoe sizes and higher reading abilities. However, in a statistical analysis that adjusts for age in some appropriate way, we expect shoe size and reading ability to be (at least approximately) independent, because we believe that there is no causal connection between them.[1]

For STR matches, the relevant extrinsic factor, the equivalent of the child's age in the example above, is the degree of relatedness between the two individuals. If i and s are directly related through one or more known common ancestors (e.g. grandparents), then matches at distinct loci are not independent, but an appropriate adjustment for the relationship can restore approximate independence (see Section 6.2.4 below). All humans are related at some level, and if i and s are apparently unrelated, this only means that the relationship between them is unknown and presumed to be distant. However, it is not necessarily distant enough to be negligible for the purposes of calculating match probabilities. You and I could well have a common ancestor as little as five generations ago: few people fully know

---

[1] If the dependence is directly causal, for example, between smoking and lung cancer, it may not be possible to eliminate the dependence by adjusting according to a third variable. This situation is central to the notion of "causality". It can be extremely difficult to convincingly establish that no such adjustment is possible.

their true ancestry that far back, and thanks to international wars, colonizations, and migrations, this possibility remains even if we were born in different parts of the world.

The question

"Is the product rule valid[2] for forensic STR loci?"

does not have a clear answer without making the question more precise. If the product rule is understood to mean multiplying together match probabilities based on an assumption of no relatedness, then the answer is a clear "no". However, the match probabilities (6.3) and (6.4) are conditional on a level of coancestry between apparently unrelated individuals measured by θ. I have argued in Chapter 5 that coancestry is the only population-genetic factor that causes a deviation away from independence systematically disfavouring defendants, and it follows that the product rule is valid given these match probability formulas and an appropriate value for θ. Use of the product rule with a θ adjustment implies an assumption of linkage equilibrium (Section 5.5) within subpopulations, but not in the broader population. The latter may manifest LD due to subdivision and drift, or admixture, that can be accounted for via θ.

Although match probabilities at many loci cannot be readily checked, the available population-genetics theory and data support the view that, given a suitable value of θ, match probabilities obtained as products of (6.3) and (6.4) will not be systematically unfavourable to defendants. If this argument is accepted, then match probabilities at multiple loci can be obtained by multiplying together the match probabilities at the individual loci calculated using (6.3) and (6.4).

Ayres (2000a) investigated the effects of inbreeding within subpopulations on two-locus match probabilities. She found that multiplying θ-adjusted single-locus match probabilities is conservative for double heterozygotes but not for double homozygotes. The overall effect is small relative to the effect of using a θ adjustment, particularly when a conservative value is used for θ. Here, and subsequently, "conservative" applies to approximations or estimates that tend to err in favour of the defendant; for a discussion, see Section 6.3.

Whole-profile match probabilities using the θ-adjusted product rule applied to the four STR loci of Table 6.1, and using various values of θ, are given in the final row of Table 6.2. Assuming $\theta = 5\%$ increases the four-locus match probability over 5-fold relative to the $\theta = 0$ case, and this extrapolates to more than a 50-fold increase for 10 loci and about 200-fold for a 13-locus profile match.

6.2.4 Relatives of s

So far, we have been considering alternative possible culprits i that are unrelated to s. Here, we consider the possibility that i and s are closely related. We continue

---

[2]Here, "valid" is understood to mean "provides a satisfactory approximation". In particular, "satisfactory" implies "not systematically biased against defendants"; some bias in favour of defendants is usually regarded as acceptable, see the discussion in Section 6.3.

IDENTIFICATION

Table 6.2 Match probabilities for the four-locus STR profile of Table 6.1 under some possible relationships of s and i, and for various values of θ.

Relationship	Match probability				
	$\theta = 0$	$\theta = 1\%$	$\theta = 2\%$	$\theta = 5\%$	
Identical twin	1	1	1	1	
Sibling	17	19	20	23	$\times 10^{-3}$
Parent/child	14	17	19	29	$\times 10^{-4}$
Half-sib	23	29	35	61	$\times 10^{-5}$
Cousin	6	8	10	20	$\times 10^{-5}$
Unrelated	8	12	17	43	$\times 10^{-6}$

Table 6.3 Distribution of ibd status and coefficient of relatedness for some possible relationships of s and i. The value of κ_j is the probability that i and s share j alleles at a locus identical by descent (ibd) from a recent common ancestor, and $\bar{\kappa}$ is half the expected number of alleles shared ibd. The values for aunt, uncle, niece, nephew, grandparent, and grandchild are the same as for half-sib.

Relationship	κ_0	κ_1	κ_2	$\bar{\kappa}$
Identical twin	0	0	1	1
Sibling	1/4	1/2	1/4	1/2
Parent/child	0	1	0	1/2
Half-sib	1/2	1/2	0	1/4
First cousin	3/4	1/4	0	1/8
Double first cousin	9/16	6/16	1/16	1/4
Unrelated	1	0	0	0

to assume that the DNA profile of i is not available to the court: it would, in principle, be desirable to exclude close relatives of s from suspicion by examining their DNA profiles, but this is rarely possible in practice. Considering first a single locus, let Z denote the number of alleles at the locus that i and s share ibd (identical by descent) from a known, recent common ancestor (e.g. parent or grandparent), and let

$$\kappa_j = P(Z = j),$$

for $j = 0, 1, 2$. The values of κ_j under some common, regular (i.e. no inbreeding) relationships are shown in the first three columns of Table 6.3.

If $Z = 0$, we are in the situation of Section 6.2.2 (s and i unrelated) and we write M_2 for the appropriate match probability, either (6.3) or (6.4). If $Z = 2$, a match is certain. For $Z = 1$, consider first the case that $s \equiv AA$. Since one allele of i is ibd with an observed allele of s, the required probability (6.2) is the probability of observing a further A allele given that two alleles have been observed to be both A. Using (5.6) with $m = n = 2$, we have

$$M_1 = P(\text{A} \mid \text{AA}) = \frac{2\theta + (1-\theta)p_\text{A}}{1+\theta}.$$

If $s \equiv \text{AB}$, the allele shared ibd by i and s is equally likely to be A or B. The match probability (6.2) is equivalent to the probability of observing the non-ibd allele given that A and B have been observed:

$$M_1 = \frac{P(\text{A} \mid \text{AB}) + P(\text{B} \mid \text{AB})}{2} = \frac{\theta + (1-\theta)(p_\text{A} + p_\text{B})/2}{1+\theta}.$$

The overall single-locus match probability for relatives is then

$$\kappa_2 + \kappa_1 M_1 + \kappa_0 M_2. \tag{6.5}$$

Four-locus match probabilities based on (6.5) and multiplication over loci are shown in Table 6.2 for the four STR loci of Table 6.1. Notice that the value of θ still affects the match probability even when i and s are assumed to be closely related. This is because, besides as matches arising via alleles shared ibd from the recent known ancestors of i and s, alleles can also be shared ibd from more distant common ancestors. The latter possibility is accounted for in the θ value. The relative importance of θ declines as the known relationship between i and s becomes closer. See Balding and Nichols (1994) for further discussion.

6.2.5 Confidence limits †

Some authors have discussed the need for confidence limits, or similar measures of uncertainty, on likelihood ratios. Forensic scientists will be aware that confidence intervals are routinely used to measure the uncertainty about an unknown parameter in many areas of science. However, note an important distinction: if you are interested in an unknown quantity that has a continuous range of possible values–the distance to the moon, say–then it makes sense to give an estimate of the uncertainty in any particular value. In the Bayesian approach, every possible value for the distance to the moon is assigned a probability (density), and these can be used to compute, say, a shortest 95% probability interval for the unknown value. The probability assignments represent uncertainty about the true distance, and it makes little, if any, sense to measure uncertainty about this uncertainty.

The probabilities making up a likelihood ratio R_i concern unknowns that have only two possible values: is s the source of the crime scene DNA (yes or no)? is i the source of the crime-scene DNA (yes or no)? Each probability represents uncertainty about the correct yes/no answer, and again there is little benefit from trying to measure the uncertainty about this uncertainty.

The relative frequency of an allele or entire profile, either the actual population proportion or the theoretical value under an evolutionary model, are unknowns for which it does make sense to give some measure of uncertainty about an estimated value, for example a confidence interval. The rationale behind confidence

IDENTIFICATION

intervals is similar to that underpinning classical hypothesis testing, to be discussed in Section 8.3. It is based on imagining a long sequence of similar "experiments". Roughly speaking, a 95% confidence interval is calculated via a rule that has the property that in 95% of repeated "experiments" the computed confidence interval would include the true value.

If relatedness and θ are ignored, we can imagine an experiment of choosing alleles at random in the population and calculate a confidence interval for a profile proportion that takes only the uncertainty due to sampling into account. This may be achieved, for example, using the asymptotic normality of its logarithm (Chakraborty *et al.* 1993). However, as noted by Curran *et al.* (2002), the approach has severe limitations in more complex scenarios, involving, for example, relatedness and mixtures. In such settings, it can be difficult to specify the imaginary sequence of similar cases.

I have argued in Section 6.2.1 that although estimates of population allele or profile proportions can be helpful in formulating the required probabilities, they are not in themselves the directly relevant quantities for assessing weight of evidence. It follows that confidence intervals for them are of little use. Buckleton *et al.* (2004) suggest that, irrespective of the philosophical arguments, a forensic scientist is likely to be asked for a confidence interval, or some similar measure of uncertainty about a reported likelihood ratio, and s/he may appear unscientific if s/he fails to provide it. This does not correspond to my own experience. There is possibly no harm in providing a confidence interval for a population profile proportion if a court calls for it, but a practice of routinely reporting such a confidence interval offends against the principle that a forensic scientist should avoid, as far as possible, presenting spurious information.

There is a legitimate concern that any calculation of a likelihood ratio requires assumptions and data, and two equally competent forensic scientists may employ (slightly) different assumptions and data and hence arrive at different likelihood ratios. A court may wish to have an indication of alternative reasonable assumptions and how big an effect they may have on the likelihood ratio. What matters most for forensic match probabilities is the level of coancestry between i and s, and uncertainty about any specific profile match probability can best be conveyed by exploring different assumptions about the value of θ. I have found that giving a range of likelihood values under different assumptions about θ satisfactorily addresses the issue of uncertainty about a particular value for the likelihood ratio.

6.2.6 Other profiled individuals

So far, we have assumed that the DNA evidence consists only of the profile of s and the crime-scene profile. In practice, however, the profiles of many other individuals are potentially relevant to the match probability.

Observed frequencies of STR alleles in population databases of anonymous individuals form an important source of information relevant to match probabilities. In the approach based on "plug-in" estimates for θ and the p_j (see page

87), these data enter the match probability only via the parameter estimates, to be discussed below in Section 6.3. We noted when introducing the plug-in approach that integrating the match probability with respect to a probability distribution for the parameters is preferable, at least in principle, though more complex. Pueschel (2001) attempts a full probability approach, explicitly conditioning the match probability on all the DNA profiles in a database. He takes population subdivision into account, but does not give any special status to the profile of s, neglecting the fact that this profile has a particular relevance to the crime that is not shared by the database profiles.

In addition to population databases of anonymous profiles, the profiles of known individuals may have been considered during the crime investigation. We assume that none of these profiles matched that of s (and hence also the crime-scene profile); otherwise, no case can proceed against s without substantial further evidence incriminating him and not the other matching individual.

The excluded individuals may have been possible suspects profiled in the course of the crime investigation, or their profiles may have already been recorded in an intelligence database. In either case, as discussed in Section 3.4.5, the effect of every excluded possible suspect is to (slightly) increase the overall case against the defendant s, since an alternative to the hypothesis that s is guilty is thereby eliminated, removing a term from the denominator of (3.3). In addition, the observation of non-matching profiles slightly strengthens the belief that the observed crime-scene profile is rare. Both these effects are typically very small and it would usually seem preferable to neglect the observed non-matching profiles when calculating match probabilities. This practice is slightly favourable to defendants and therefore need not be reported to the court, nor any adjustment need be made to the likelihood ratio because of it.

6.3 Application to STR profiles

According to the weight-of-evidence formula (3.3), a rational juror should assess a likelihood ratio for every alternative possible culprit i. Under the assumptions of Section 6.2, these likelihood ratios are equivalent to match probabilities, and we have derived formulas for them in terms of population allele proportions (the p_j) and θ. Instead of integrating the match probability over these unknown values, with respect to a probability distribution representing uncertainty about their values for the alternative suspects in a particular case, we advocate here a simpler approximation obtained by "plugging in" estimates for these parameters. We now consider the choice of plug-in values.

One possible reason for a forensic scientist to use relatively large values for θ and the p_j is a desire on the part of courts to be conservative (tend to favour the defendant). Excessive conservatism could lead to an unnecessary and undesirable failure to convict the guilty, and it is difficult to formulate principles for how much conservatism is appropriate. Despite this difficulty, it does seem

IDENTIFICATION

worthwhile, at least for criminal cases, to adhere to the principle that the prosecution should not, as far as reasonably possible, overstate its case against a defendant.

6.3.1 Values for the p_j

In typical forensic identification problems, we have available estimates of allele proportions in broadly defined population groups, such as Caucasians, sub-Saharan Africans, East Asians, or Aboriginals. Recall that in the single-contributor identification setting, the match probability (6.2) is the conditional probability that i has profile D given that s has it. It follows that, in principle, the population database to be used to estimate the p_j should be that most appropriate for i, the alternative possible culprit under consideration. It is simpler, and usually conservative, to use the database closest to the suspect s for all i, together with an appropriate value for θ. There is sometimes difficulty in choosing the database most appropriate for s, for example, because s/he is of mixed race or from an admixed population, or a minority group, or simply because the population from which the database is drawn is not well defined. If so, it may be acceptable to use the database yielding the largest match probability. The additional uncertainty arising in such cases can be allowed for in choosing the value of θ: the "worse" the database is in reflecting the genetic background of a possible culprit, the greater is the appropriate value of θ.

Forensic population databases usually consist of at least several hundred profiles. This is large enough that allele proportions above 1% can be estimated reasonably well, provided that the sample is representative (see Section 6.3.2 below). However, there remains some sampling uncertainty, whose effects on match probabilities need to be accounted for (Section 2.3.1).

One approach to accounting for the sampling uncertainty in observed allele proportions is to estimate p_j at a heterozygous locus by

$$\widehat{p}_j = \frac{x_j + 2}{n + 4}, \qquad (6.6)$$

where x_j is the frequency of allele j in the database, and n is the total frequency of all alleles: $n = \sum_j x_j$ (see Balding 1995; Balding and Nichols 1994). In the homozygous case, an analogous estimate is

$$\widehat{p}_j = \frac{x_j + 4}{n + 4}. \qquad (6.7)$$

These estimates can be thought of as adding both the crime-scene and defendant profiles to the database; such estimation methods are sometimes called "pseudo-count" methods. They can be justified as approximations to the posterior mean given the sample allele counts and a (multivariate) uniform prior distribution[3] for

---
[3] More generally, a Dirichlet distribution.

the allele proportions. The effects of using these estimates, rather than the database proportions x_j/n, are

(i) smoothing out some sampling variation for low-frequency alleles, and

(ii) building in some conservatism by increasing the estimates of the p_j for the observed alleles, by a relatively large amount when they are rare.

As a direct consequence of (ii), the allele proportions used in different cases sum to more than one. This property also applies to the use of upper confidence limits, or any other conservative estimators, in place of the p_j.

Curran *et al.* (2002) are critical of the approach based on using the "plug-in" estimates (6.6) and (6.7) in place of a more careful accounting for the sampling uncertainty in estimates of the p_j. I acknowledge that it is a rough rule-of-thumb approach. It is not intended to serve as an approximate confidence limit, nor indeed to satisfy any specific criterion of rationality. I believe that Curran *et al.* will agree with me that the "gold standard" is to integrate over a full probability distribution for the p_j given the database frequencies and a reasonable prior distribution. However, this can be computationally intensive for little gain, and there are interpretational advantages in using explicit values for the p_j. Moreover, uncertainty about the p_j is in practice relatively unimportant compared with the effects of the value of θ.

6.3.2 The value of θ

In Section 5.2, we defined θ in terms of the mean square error of the allele proportion in a subpopulation to which i and s both belong, relative to a reference value such as the allele proportion in the population from which the available database was drawn. In Section 5.6, we showed that the value for θ can also be interpreted in terms of the coancestry of i with s, and in Section 5.7, we showed that estimates of θ can depend sensitively on the choice of reference values.

Published estimates of θ at STR loci for subpopulations within the major human "racial" groups are often small, and typically less than 1%. See, for example, Ayres *et al.* (2002) for some θ value in Australian populations of Caucasians, Asians, and Aboriginals, and Balding *et al.* (1996) and Balding and Nichols (1997) for estimates in some European Caucasian populations. Many other authors have estimated θ at forensic STR loci in various populations and using different methodologies; see Buckleton *et al.* (2004) for a brief review. Typically, they report even smaller values than those cited above.

There are several arguments for using larger values in forensic practice than suggested by these estimates, beyond the desire for conservativeness discussed briefly above. Recall that our goal is not to estimate a "best" value for θ but to choose "plug-in" values that give a match probability similar to that which would be obtained from using the full Bayesian approach.

IDENTIFICATION

- Because of the skewness of appropriate distributions for θ, the "plug-in" value that mimics the effect of integration over θ may be much larger than, say, a maximum-likelihood estimate of θ.

- The expected value of θ^2 can be much larger than the square of the expected value of θ, and this effect is stronger for higher powers. Since multi-locus match probabilities involve high powers of θ, plug-in values should tend to be towards the higher end of the range of plausible values to allow for this effect.

- Published estimates of θ usually relate to the variation of allele proportions around the observed mean of the subpopulations studied. However, in forensic applications, the reference value is the forensic database value, and variation about this value may well be substantially larger than that about a mean value. For example, in Table 5.3, the θ estimate varies substantially according to the choice of reference value and is often much larger than when the reference value is estimated from the data. Further, minority "racial" groups may be heterogeneous, and the database allele proportion may not be representative of the specific ethnic group relevant to a particular crime. More generally, forensic databases are typically not scientific, random samples but are "convenience" samples whose representativeness and the magnitude of the resulting uncertainty in allele proportion estimates are difficult to assess precisely. However, its effects can be allowed for by using an appropriately large value of θ.

- Although we introduced θ in terms of the variance of subpopulation allele proportions, subpopulations are a theoretical construct and difficult to specify precisely. (In the United Kingdom, do Catholics form a subpopulation? do the people of Northumberland form a subpopulation? university graduates? what about Catholic, university graduates from Northumberland? In each case, some tendency to mate within the group is plausible). The coancestry interpretation of θ (Section 5.6) is in terms of the shared ancestry of pairs of individuals: the individuals i with larger values of θ with respect to s will contribute disproportionately to the total weight of evidence in formula (3.3), and using an "average" value for θ will not reflect this.

For these reasons, I suggest that a relatively large value, such as 2%, be used when both suspect and alternative possible culprit are drawn from a relatively well-mixed, large population, and perhaps 3% could be used if both are drawn from one of the large minority groups. In some small minority groups, $\theta = 5\%$ may be appropriate. These suggested values are based on informal judgements, and it is difficult to justify more precise values.

If i is not from the same "racial" group as s, then they have little coancestry, and so a small value of θ can be justified. However, since the problem of representativeness of any database remains, as well as possibilities for some coancestry even across apparent racial groups, I advocate using a non-zero value of θ in every case, perhaps setting 1% as the minimum.

6.3.3 Errors

We saw in Section 4.1.1 that there are a number of factors that can lead to false exclusion error due to a true culprit's STR profile being reported incorrectly from either the crime or evidence sample. The overall false exclusion rate is thought to be extremely small and is negligible compared with the rate of unsolved crimes. In any case, we saw in Section 2.3.3 that, having observed a match, the possibility of a false mismatch is almost irrelevant to evidential weight.

Potentially much more important are false inclusion errors, in which an innocent individual is reported to have a profile that does not exclude him from being the culprit. As we noted in Section 3.4.4, there are at least two ways in which this can occur: a chance match, the probability of which is assessed by the match probabilities calculated in Section 6.2, or a false match, perhaps due to an incorrect recording of profiles from either crime or evidence samples. We also noted in Section 3.4.4 that

(i) the match probability is effectively irrelevant if swamped by the probability of a false-match error;

(ii) it is not the probability of any error that matters but only an error that leads to a false inclusion.

Some critics of the reporting of DNA profile matches have argued that (i) above is routinely the case, and hence the match probabilities reported in court are irrelevant and potentially misleading. This argument has some force, and protagonists of DNA profiling have been prone to mistakes and exaggeration in their attempts to discount it (see, for example, Koehler 1996). However, because of (ii), the chance of a false inclusion error due to genotyping anomalies is so remote as to be negligible, even relative to the match probability. Contamination is a distinct possibility in some settings, but because evidence and crime samples are routinely typed by different staff in different laboratories, sometimes with a substantial time gap, in many cases this possibility can also be ruled out reasonably.

There remains the possibility of false inclusion due to handling or labelling error, or evidence tampering or fraud such as a "frame-up". A conspiracy theory, such as that police and/or judicial authorities colluded to manufacture evidence falsely linking the defendant with the crime scene, may be regarded by a reasonable juror as substantially more plausible than a chance match whose probability is reported as less than 1 in a billion, even when there are no particular reasons to suggest a conspiracy.[4] The likelihood ratio for the DNA evidence under such a conspiracy theory is likely to be close to one, so that the juror may as well dismiss the match probability as irrelevant (see Exercise 2, Section 6.6). The relevant probability for the juror to assess is the probability of such a conspiracy based on all the other evidence.

---

[4]The case of OJ Simpson (California, 1995) provides an example in which the defence argued that there was evidence of evidence tampering, and the jury acquitted. See the discussion in Buckleton *et al.* (2004).

IDENTIFICATION

There seems no role for a forensic scientist to predict whether a juror might pursue such a line of reasoning, and hence the only reasonable option is to supply the juror with a match probability but also to try convey an understanding of the circumstances under which the match probability would be effectively irrelevant.

6.4 Application to haploid profiles

6.4.1 mtDNA profiles

Profiles based on mtDNA sequences were introduced in Section 4.2. Because of the absence of recombination, the whole mtDNA sequence can be considered as a single locus with many possible alleles. Unlike autosomal loci, at which we carry two alleles, we normally carry just one mtDNA allele.

The uniparental inheritance of mtDNA has advantages in establishing the relatedness of individuals through female lines: close maternal relatives of an individual will have (almost) the same mtDNA. This feature was used to help identify the remains of the Russian royal family (Gill *et al.* 1994), in part using mtDNA from a member of the British royal family. However, this feature can be a disadvantage for forensic identification because an individual cannot easily be distinguished from any of his/her maternal relatives, even those removed by several generations. Moreover, a person's maternal relatives are likely to be unevenly distributed geographically, and it will often be difficult to assess how many close maternal relatives of s should be considered as alternative possible sources of an evidence sample. The high mutation rate of many mtDNA sites lessens this problem but does not eliminate it. On the other hand, the high mutation rate leads to the problem of heteroplasmy (multiple mtDNA types in the same individual). As noted in Section 4.2, heteroplasmy can be problematic for inference because not all of an individual's mtDNA sequences may be observed in, for example, typing a small and/or degraded crime-scene sample.

Currently, the most widespread approach to interpreting mtDNA profiles in an identification problem in which there is a match between suspect and crime-scene mitotypes ($\equiv$ mtDNA types) is to report this evidence along with the frequency of the mitotype in a population database of size N. This simple approach fails to deal adequately with the complications outlined above, as we now briefly discuss.

To make some allowance for sampling variability, it is advantageous to include both the crime scene and the defendant profiles with those of the population database. For autosomal markers, this practice leads to the "pseudo-count" estimates (6.6) and (6.7). Here, we obtain

$$\widehat{p} = \frac{x+2}{N+2}, \tag{6.8}$$

where x denotes the database count of the mitotype. Wilson *et al.* (2003) considered a sophisticated genealogical model that could exploit the sample frequencies of similar profiles to refine estimates of p, but they found that $\widehat{p}$ given at (6.8) provided a satisfactory, simpler estimator that is typically slightly conservative.

The above analyses take no account of the effects of any population substructure. Using an approach analogous to that used to derive (6.3) and (6.4), the match probability for an mtDNA allele with population frequency p is obtained by dividing the case $m = n = 2$ of the sampling formula (5.11) by the case $m = n = 1$, which gives:

Match probability: haploid case

$$R_i = \theta + (1 - \theta)p. \tag{6.9}$$

Since p is typically unknown, (6.9) can be combined with (6.8) to obtain a practical formula. This gives essentially the formula reported as an "intuitive guess" by Buckleton et al. (2004).

Formula (6.9) applies to a structured population in which the subpopulations are well mixed. It does not take account of the fact that maternally related individuals might be expected to be tightly clustered, possibly on a fine geographical scale. Reports of θ estimates for mtDNA drawn from cosmopolitan European populations typically cite low values, reflecting the fact that this population is reasonably well mixed, as well as the effects of high mtDNA mutation rates. However, researchers rarely are able to focus on the fine geographic scale that may be relevant in forensic work, and there are some large θ estimates at this scale. See Buckleton et al. (2004) for a discussion of mtDNA θ estimates and their limitations.

In the presence of a matching heteroplasmic profile, no modification to (6.9) is required except that p refers to the population proportion of individuals displaying the same heteroplasmy. The value of p will thus be difficult to estimate and will, in part, depend on the typing platform employed. Imperfect matches can arise with heteroplasmic profiles. For example, the crime-scene mtDNA profile may consist of just one sequence, say d, whereas the mtDNA profile of s displays a set of sequences D that includes d. In this case, the denominator of the likelihood ratio (3.1) requires an assessment of the probability that, under the circumstances of the recovery of crime-scene DNA, only d would be recorded rather than the full, heteroplasmic profile D if s is the culprit. The numerator of the likelihood ratio (i is the culprit) requires an estimate of the joint probability of the two observed mitotypes given that i and s are distinct (but possibly related) individuals.

Although selection is thought likely to influence the distribution of mtDNA types, because no assumption of either Hardy-Weinberg or linkage equilibrium is made in calculating mtDNA match probabilities, it seems reasonable to assume that selection will not adversely affect the validity of (6.9).

IDENTIFICATION

6.4.2 Y-chromosome markers

See Section 4.3 for a brief introduction to the properties of Y-chromosome markers.

The interpretation issues for Y-chromosome profiles are very similar to those for mtDNA. In particular, (6.8) and (6.9) also apply. Roewer *et al.* (2000) advocated a method of estimation of the population proportion of Y haplotypes that is based on the beta distribution, but which is similar in form to the pseudo-count estimator (6.8).

The problem of heteroplasmy rarely, if ever, arises for Y chromosomes, but the problem of population structure is even more important (Jobling *et al.* 1996) since values of θ are typically higher than for mtDNA. Zerjal *et al.* (2003) report a specific Y haplotype that is frequent in many parts of Asia, and largely unobserved elsewhere, and which they suggest can be attributed to the consequences of historic Mongol conquests, possibly even to Genghis Khan himself. Similar historical events on a smaller scale may have led to an unrecognized, local elevation of the concentration of a specific Y haplotype that is otherwise rare.

Micro-geographical variation in forensic Y-chromosome haplotypes has been studied by Zarrabeitia *et al.* (2003) in the Cantabria region of Spain. These authors found that substantial overstatement of evidential strength frequently results from the use of population databases collected on too broad a geographical scale.

6.5 Mixtures

6.5.1 Visual interpretation of mixed profiles

A mixed profile arises when two or more individuals contribute DNA to a sample. Torres *et al.* (2003) give a survey of mixtures that have arisen in their own casework. An example of an EPG corresponding to a mixed STR profile is shown in Figure 6.2. The profile at the Amelogenin sex-distinguishing locus (leftmost on the second, "Green" panel) shows a predominance of X, but some trace of Y, suggesting that the mixture stain may have come predominantly from a female, with a minor contribution from a male. This mixture was created in the laboratory, with known contributors, and the above interpretation is indeed correct, but let us proceed for the moment as if we did not know this.

The recorded signals at other loci are consistent with there being two contributors, one "major" and one "minor". For example, at locus VWA (second locus from the left in the top, "Blue" panel) the EPG indicates the presence of four alleles, two of which (labelled 14 and 17) produce a strong signal, while the remaining two signals (corresponding to alleles 18 and 19) are much weaker, though still clearly distinguishable from the background noise.

Two other loci show similar four-allele patterns in the EPG. However, at most of the loci, only two or three alleles appear in the EPG, which can arise if one or both contributors are homozygous, or if they have alleles in common. At locus D3 (leftmost in the "Blue" panel), there is a strong signal at allele 15, a slightly weaker

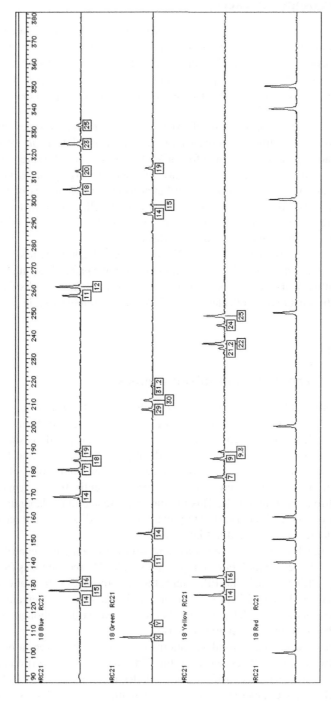

Figure 6.2 An EPG showing a mixed STR profile obtained from a DNA sample with two contributors. The major profile is from a female and the minor component from a male. The DNA samples were amplified with the SGM Plus®STR kit. The amplified fragments were separated on the ABI Prism 377 DNA Sequencer and analysed using the GeneScan®3.1.2 and Genotyper®2.5 software. Image supplied courtesy of LGC. © 2004 LGC.

signal at allele 16, and a much weaker signal at allele 14, suggesting that the major contributor has genotype 15, 16 at this locus, while the minor contributor is 14, 15. An alternative possibility is that the peak corresponding to allele 14 corresponds to a stutter peak, and only alleles 15 and 16 are actually present in the sample, but this is unlikely because the peak, although low, is higher than normal for a stutter peak at this position on the EPG. Although the major ≡ 15, 16 and minor ≡ 14, 15 genotype designations seem the most plausible, it is difficult to assign a measure of confidence to this call.

Interpretation at the D16 locus (third from left in the "Blue" panel) is even more difficult: the two observed alleles, 11 and 12, display allele signals of noticeably different heights, yet they are not extremely asymmetric. Perhaps the major contributor is 11, 12 and the minor is a 12, 12 homozygote, but other possibilities seem to exist, such as that both contributors are 11, 12 and the apparent peak imbalance results from a fluctuation in the experimental conditions: here the ratio of peak heights is about 0·72, which may be regarded as within the normal range of variation of peak heights for a single, heterozygous contributor.

As the above discussion suggests, inferring the profiles of major and minor contributors to a sample can sometimes be done with reasonable confidence, but often it is problematic at least for some loci. The presence of good-quality DNA and a strong imbalance in the proportions of the DNA from each source individual facilitate the task. However, in the presence of degraded samples, low DNA copynumber, an unknown number of contributors, or an equal contribution from two contributors, the task can be challenging. Although automated procedures exist, in the presence of the complications listed above, these seem not yet to be sufficiently reliable that they can replace visual interpretation. Most importantly, assigning a measure of confidence to any particular genotype designation is problematic.

6.5.2 Likelihood ratios under qualitative interpretation

One approach to overcoming the problems with visual interpretation of mixtures, at the cost of discarding the quantitative information from the EPG, for example, about peak heights and shapes, is to limit interpretation to qualitative allele calling only, without any attempt to infer the underlying genotypes. Thus, the interpretation of locus VWA in the EPG of Figure 6.2 would be limited to the conclusion that alleles 14, 15, and 16 are observed in the mixture. Then, all combinations of underlying genotypes that includes at least one copy of each of these alleles are regarded as equally plausible.

We consider here the single-locus case; likelihood ratios can be combined across loci via multiplication (see Section 6.2.3). These approaches were initially developed by Evett *et al.* (1991) for non-STR profiles, but they remain applicable to STR profiles. Mortera *et al.* (2003) describe a probabilistic expert system (Section 7.3) for the qualitative analysis of DNA mixtures, and the qualitative approach has recently been extended to take account of the coancestry of all the contributors to a mixture (Curran *et al.* 1999; Fung and Hu 2000, 2002) and alternative suspects related to the accused (Fung and Hu 2004). These authors also offer software for mixture interpretation (Hu and Fung 2003).

The number of unknown contributors

For SNP profiles (Section 4.5), inferring the number of contributors to a profile is always problematic since no more than two alleles are recorded (Egeland *et al.* 2003). For the STR profiles that are currently predominant in forensic work, a minimum number of contributors to a sample of DNA is provided by half the number of distinct alleles observed at any one locus. Even in the case that no more than two alleles are observed at any locus, the usual assumption of only one contributor to the sample is an assumption: it is not implied by the evidence. The probability of observing no more than two alleles at any locus, given multiple sources, is typically extremely small. Thus, absent an extremely strong prior belief that there were multiple sources, a single source is very likely on the basis of the evidence.

In principle, prior to the DNA evidence, judgements about the number of possible sources should lie in the domain of the court, not in that of the DNA expert. There will inevitably be occasions when experts make such prior judgements, when they seem uncontroversial, and when the alternative may be to overly complicate their evidence (e.g. by working out likelihood ratios under many different scenarios, most of which are implausible). However, the problem of the number of contributors to a mixture can sometimes be overcome without an inappropriate assumption by the forensic scientist by finding appropriate bounds on the likelihood ratio; see for example Buckleton *et al.* (1998) and Lauritzen and Mortera (2002).

Two contributors: one known, one unknown

The easiest case arises when all but one of the contributors of a mixture are individuals whose DNA profiles are known and who are not considered as possible culprits, for example, victims or bystanders. If there is one known contributor, v, and s is accused of being the (unique) unknown contributor, then the likelihood ratio takes the form

$$R_i = \frac{P(E \mid i \text{ and } v \text{ are the sources of the crime-scene DNA})}{P(E \mid s \text{ and } v \text{ are the sources of the crime-scene DNA})}. \qquad (6.10)$$

The DNA evidence E consists of all three observed profiles, those of s, v, and the crime stain. If we assume that the probabilities that s and v have particular profiles are the same whether or not they are the contributors to the crime-scene profile, then (6.10) can be expressed as

$$R_i = \frac{P(\text{crime-scene profile} \mid \text{profiles of } s \text{ and } v; \text{ sources: } i \text{ and } v)}{P(\text{crime-scene profile} \mid \text{profiles of } s \text{ and } v; \text{ sources: } s \text{ and } v)}. \qquad (6.11)$$

IDENTIFICATION

Example

Consider a single locus, and suppose that the following alleles are observed:

$$\text{defendant } s \equiv AB$$

$$\text{victim } v \equiv AC$$

$$\text{crime stain} \equiv ABC$$

in which we use $\equiv$ to denote "has genotype". Then (6.11) becomes:

$$R_i = \frac{P(ABC \mid s \equiv AB, v \equiv AC; \text{sources: } i, v)}{P(ABC \mid s \equiv AB, v \equiv AC; \text{sources: } s, v)}. \qquad (6.12)$$

The denominator equals one. For the numerator, given that $v \equiv AC$, the crime-scene genotype implies that the genotype of i

- includes a B allele, and
- does not include any allele other than A, B, and C.

The possible genotypes consistent with these two requirements are AB, BB, and BC. Thus, the numerator of (6.12) is the probability that either $i \equiv AB$, $i \equiv BB$, or $i \equiv AC$ given that $s \equiv AB$ and $v \equiv AC$. We assume

- that s, v, and i are mutually unrelated;
- no coancestry (i.e. $\theta = 0$);
- that the allele proportions p_A, p_B, and p_C are known; and
- genotypes are in Hardy-Weinberg proportions (see Section 5.4).

Then the genotypes of s and v have no effect on the numerator, and (6.12) becomes:

$$R_i = P(i \equiv AB) + P(i \equiv BB) + P(i \equiv BC)$$

$$= 2p_A p_B + p_B^2 + 2p_B p_C = p_B(2p_A + p_B + 2p_C)$$

This calculation ignores possible coancestry between i and one or both of s and v. If i, s, and v are all assumed to have a common level of coancestry measured by θ, then we need to take into account the four alleles (AABC) already observed in s and v, so that

$$R_i = 2P(AB \mid AABC) + P(BB \mid AABC) + 2P(BC \mid AABC). \qquad (6.13)$$

Just as for the derivations of (6.3) and (6.4), the conditional probabilities of (6.13) can be evaluated by two applications of the recursive form of the sampling formula, (5.6). For example $P(AB \mid AABC)$ is the probability of observing an A followed by a B in two further draws from a population when a sample of size four has already been observed to be AABC. This probability can be computed as the product of instances of (5.6) with $m = 2$ and $n = 4$, and with $m = 1$ and $n = 5$. Working similarly for the other two terms, we obtain

$$R_i = 2\frac{(2\theta + (1-\theta)p_A)(\theta + (1-\theta)p_B)}{(1+3\theta)(1+4\theta)}$$

$$+ \frac{(2\theta + (1-\theta)p_B)(\theta + (1-\theta)p_B)}{(1+3\theta)(1+4\theta)}$$

$$+ 2\frac{(\theta + (1-\theta)p_B)(\theta + (1-\theta)p_C)}{(1+3\theta)(1+4\theta)}$$

$$= \frac{(\theta + (1-\theta)p_B)(8\theta + (1-\theta)(2p_A + p_B + 2p_C))}{(1+3\theta)(1+4\theta)}$$

All three possibilities for the genotype of an alternative contributor, i, involve alleles that have already been observed in the genotypes of either s or v. Thus, possible coancestry has the effect of increasing[5] the probability of all possible genotypes for i over the $\theta = 0$ case.

In the above example, the crime-scene genotype includes one allele not shared with v, and hence this must have come from the other contributor. There are essentially only two other cases: that zero and two alleles, respectively, from the unknown contributor can be determined by subtracting v's alleles from the crime-scene genotype. An example of the former situation arises when both the crime-scene genotype and the genotype of v are of the form AB. Then the genotype of the unknown contributor can be any of AA, AB, or BB. If instead, the crime-scene genotype is ABC and that of v is AA, then the genotype of the unknown contributor must be BC.

Two unknown contributors

Here, difficulties arise because of the number of scenarios to explain the components of the mixture. For example, consider the case of two co-defendants, s_1 and s_2, both accused of raping the same victim. Suppose that the DNA genotype obtained from crime-scene semen matches the combined genotypes of s_1 and s_2, and that the victim's DNA is not included in the crime stain.

In assessing the case against s_1, three distinct types of likelihood ratio may be considered:

$$R_i^a = \frac{P(E \mid i_1 \text{ and } s_2 \text{ are the sources})}{P(E \mid s_1 \text{ and } s_2 \text{ are the sources})}$$

$$R_i^b = \frac{P(E \mid i_1 \text{ and } i_2 \text{ are the sources})}{P(E \mid s_1 \text{ and } s_2 \text{ are the sources})}$$

$$R_i^c = \frac{P(E \mid i_1 \text{ and } i_2 \text{ are the sources})}{P(E \mid s_1 \text{ and } i_1 \text{ are the sources})}$$

---

[5] The probability could decrease if one of p_A, p_B, or p_C were much larger than 0.5.

IDENTIFICATION

where i_1 and i_2 are alternative possible culprits.

- R_i^a is useful for assessing the case against s_1 on the hypothesis that s_2 is a contributor to the crime stain. Calculation of this likelihood ratio is the same as in the case of one unknown contributor considered above. However, use of R_i^a in the trial would typically amount to an assumption of the guilt of s_2 in assessing the case against s_1, and this would not usually be appropriate when they are co-defendants being tried simultaneously, or for the first defendant if they are being tried sequentially.

- R_i^b is useful for assessing the case against s_1 and s_2 jointly, and so it does not respond to the juror's need, under most legal systems, to make separate decisions about the guilt of s_1 and s_2. The juror must assess the individual cases against each man, and R_i^b is not appropriate for this.

- R_i^c can be used to assess the case against s_1 under the assumption that the second contributor to the crime stain is unknown, and will typically be most appropriate form of likelihood ratio for assessing the evidence against s_1, without making an assumption about whether s_2 is a contributor.

Whichever likelihood ratio is used, its assumptions need to be clarified to the court. We will henceforth only consider the likelihood ratio R_i^c, and drop the superscript c from the notation.

Example
Suppose that, at a particular locus, the following genotypes are observed:

$$s_1 \equiv AB$$

$$s_2 \equiv CC$$

$$\text{crime scene} \equiv ABC$$

The crime-scene genotype is consistent with the contributors having genotypes AA and BC, for example, or AB and BC, in addition to being consistent with s_1 and s_2 being the sources of the DNA. Then,

$$R_i = \frac{P(ABC \mid s_1 \equiv AB, s_2 \equiv CC, i_1 \text{ and } i_2 \text{ are the sources})}{P(ABC \mid s_1 \equiv AB, s_2 \equiv CC, s_1 \text{ and } i_1 \text{ are the sources})}. \quad (6.14)$$

Its evaluation will depend on the relationships among s_1, s_2, i_1, and i_2. Here, for simplicity, we will assume that all are unrelated, and initially we will also set $\theta = 0$. Under these assumptions, the genotype of s_2 is irrelevant to R_i.

The numerator of (6.14) is $12p_A p_B p_C(p_A + p_B + p_C)$. To see this, consider the event that it is the C allele that arises twice in the genotypes of i_1 and i_2: this could result from a CC homozygote and an AB heterozygote (probability $4p_A p_B p_C^2$; one factor of two comes from the heterozygote, the other from the orderings of the two genotypes), or from an AC heterozygote and a BC heterozygote (probability $8p_A p_B p_C^2$), giving a total probability of $12p_A p_B p_C^2$. The expression for the numerator comes from combining this with the two terms corresponding to the A and the B alleles being represented twice.

The denominator of (6.14) is the probability that i_1 has genotype AC, BC, or CC, which is $p_C(2p_A + 2p_B + p_C)$, and so

$$R_i = \frac{12 p_A p_B (p_A + p_B + p_C)}{2p_A + 2p_B + p_C}.$$

If $p_A = p_B = p_C = p$, then R_i reduces to $36p^2/5$, which takes maximum value 0.8 when $p = 1/3$: in this case the evidence is of little value; however, if p is small, the evidence is stronger: $R_i = 0.072$ if $p = 0.1$, and $R_i = 0.018$ if $p = 0.05$.

Now, assume that s_1, s_2, i_1, and i_2 are all unrelated but are drawn from the same subpopulation for which the level of coancestry, relative to the population allele frequencies, can be characterized by a given value of θ. Now, the probability that it is the C allele that arises twice in the genotypes of i_1 and i_2, given the observed genotypes of s_1 and s_2, is 12 times the probability that an ordered sample of size four is ABCC, given that a sample of size four has already been observed to be ABCC. Using the recursive form of the sampling formula (5.6) four times leads to

$$12 \frac{(\theta + (1-\theta)p_A)(\theta + (1-\theta)p_B)(2\theta + (1-\theta)p_C)(3\theta + (1-\theta)p_C)}{(1+3\theta)(1+4\theta)(1+5\theta)(1+6\theta)}$$

and the remaining two terms of the numerator may be computed similarly. The denominator of R_i becomes

$$2\frac{(2\theta + (1-\theta)(p_A + p_B))(\theta + (1-\theta)p_C)}{(1+3\theta)(1+4\theta)} + \frac{(2\theta + (1-\theta)p_C)(3\theta + (1-\theta)p_C)}{(1+3\theta)(1+4\theta)}.$$

The final result for R_i is tedious to write down but very easy to compute. For example, when $p_A = p_B = p_C = 1/3$, the value of 0.8 at $\theta = 0$ increases only slightly to a maximum of 0.8007 at $\theta \approx 1\%$, and then declines as θ increases. Thus, the assumption of a large level of coancestry among the two defendants and the alternative possible culprits in this setting *strengthens* the case against s_1. However, if $p_A = p_B = p_C = 0.1$, then increasing θ weakens the case against s_1: R_i increases from 0.072 at $\theta = 0$ to 0.094 at $\theta = 1\%$ and 0.17 at $\theta = 5\%$.

6.5.3 Quantitative interpretation of mixtures

Clayton et al. (1998) proposed a set of procedures for interpreting mixtures with two contributors that involves using peak heights from the EPG to estimate the proportion of the mixture attributable to each contributor. Possible genotype configurations that are extremely inconsistent with these mixture proportions are then excluded from consideration. For example, for the EPG of Figure 6.2 we noted in Section 6.5.1 that the peak heights at several loci indicated four distinct alleles, of which two, say alleles A and B, gave strong signals (with approximately equal

IDENTIFICATION 109

peak heights) and two, say C and D, gave weak signals (also with approximately equal peak heights). The qualitative approach discussed above would count all three genotype pairs consistent with these alleles (AB,CD; AC,BD; and AD,BC) as equally plausible. The semi-quantitative approach of Clayton *et al.* (1998) would regard only the AB,CD genotype pair as plausible because it pairs up the alleles with the two strong and the two weak signals.

This semi-quantitative approach seems to be widely employed in practice, but it is not without difficulties. Estimating mixture proportion relies on an assumption that this proportion is approximately the same across loci. This seems a reasonable assumption prior to the PCR step in the STR typing procedure, and indeed Gill *et al.* (1998) examined the estimation of mixture contribution proportions and found that consistency across loci is the norm. However, it remains possible that the PCR reaction proceeds abnormally at one or more loci, so that any such loci may have very different mixture proportions from the remaining loci, which can result in invalid inferences about plausible genotype pairs at these loci. Further, deciding which genotype configurations are consistent with the mixture proportions is always difficult in some borderline cases. For further description and criticisms, see Buckleton *et al.* (2004).

Gill *et al.* (1998) went on to suggest a more quantitative approach in which a weight was given to each possible genotype allocation according to its plausibility given the estimated mixture proportion. The ideal, fully quantitative approach to assessing weight of evidence for STR mixtures would involve all the information contained in the EPG. Evett *et al.* (1998) set out a framework for analyses taking peak areas into account. However, there are a number of difficulties in taking this approach in practice, in particular that it requires detailed modelling assumptions that would be difficult to justify. It seems not to be widely implemented as yet, but developments in this area are ongoing.

6.6 Identification exercises

Solutions start on page 166.

1. Suppose that a man s is accused of being the source of a crime sample. At two STR loci his genotypes are of the form AB and CC, respectively, in both cases matching the crime-scene genotype. The crime occurred in an isolated village and it is assumed that one of 45 men is the culprit; these include s, his half-brother h, two uncles, and 41 unrelated men. There has been no recent migration into the village, and the background level of coancestry among the men, relative to the wider population, is taken to be $\theta = 3\%$.

    (a) Calculate the probability that s is the true culprit if all 45 men are initially regarded as equally under suspicion, and the allele proportions in the wider population are $p_A = 6\%$, $p_B = 14\%$, and $p_C = 4\%$.

    (b) Repeat (a), but now there is a camp of migrant labourers nearby and 20 men from this camp are under suspicion, equally with the 45 men in the village. The migrants come from a distant population in which $p_A = 8\%$, $p_B = 9\%$, and $p_C = 6\%$.

2. †Suppose that you are the juror in a case that rests on the identification of a single contributor to a crime stain via a DNA profile match. You have heard a forensic scientist report a match probability of 10^{-9} for an alternative culprit unrelated to s. You accept this value, and also that all relatives of s are excluded from suspicion. However, you mistrust the island police and, on the basis of the non-DNA evidence, you assess that there is a 1% chance that the police have maliciously swapped the crime sample with a sample taken directly from s. How would you calculate the appropriate likelihood ratio, taking into account both the possibility of a chance match and the possibility of fraudulent evidence?

3. Using the qualitative approach of Section 6.5.2, formulate single-locus likelihood ratios conveying the weight of mixed DNA profile evidence against a suspect s in the following situations. Assume that there are exactly two contributors to the mixed profile and that all actual and possible contributors are unrelated but come from the same subpopulation characterized by a given value of θ. Express your answer in terms of θ and allele proportions such as p_A and p_B.

   (a) Only two alleles are observed in the mixed profile, say A and B, and s is an AA homozygote. One of the contributors to the profile is known to be the victim, who is an AB heterozygote at that locus.

   (b) Four alleles are observed, A, B, C, and D, and s is an AB heterozygote. One of the contributors to the profile is known to be the victim, who is a CD heterozygote at that locus.

   (c) Four alleles are observed, A, B, C, and D, and s is an AB heterozygote. The second contributor is unknown.

4. Consider again the example introduced on page 105 and assume that $\theta = 0$.

   (a) Evaluate a likelihood ratio assessing the strength of the evidence against s_2.

   (b) Suppose now that the EPG at every locus is consistent with there being a strong imbalance in the contributions to the mixture from the two source individuals. At the locus in question, the peaks corresponding to alleles A and B are about the same height, whereas the peak for allele C is much lower.

      i. What is the most likely genotype for the major contributor at this locus? Assuming this genotype, derive an appropriate likelihood ratio assessing the hypothesis that s_1 is a contributor to the profile.

      ii. What genotypes are possible for the minor contributor to the mixed profile? Derive an appropriate likelihood ratio assessing the hypothesis that s_2 is a contributor to the profile.

7

Relatedness

7.1 Paternity

7.1.1 Weight of evidence for paternity

The likelihood ratios for assessing an allegation that s is the father of a child c have the form

$$R_i = \frac{P(E \mid i \text{ is the father of } c)}{P(E \mid s \text{ is the father of } c)}, \qquad (7.1)$$

for each alternative possible father, i. In (7.1), E denotes all the evidence, but here we will ignore the non-DNA evidence (the principles are the same as in Section 3.4.2) and assume that E includes only the DNA evidence. Typically, this will consist of the profiles of the mother m, in addition to those of s and c, in which case

$$R_i = \frac{P(\text{profiles of } c, s, \text{ and } m \mid i \text{ is father})}{P(\text{profiles of } c, s, \text{ and } m \mid s \text{ is father})}. \qquad (7.2)$$

Here, the fact that m is the mother of c is regarded as background information in both probabilities, but is not explicit in the notation.

The general issues for interpretation of R_i in the paternity setting are the same as those for identification, discussed above in Sections 3.2 and 3.3. In particular, the likelihood ratios for DNA evidence can be combined with those for other evidence, for example, the statements of m and s and information about their locations of residence and the frequency of contact between them. These likelihood ratios can then be combined using the weight-of-evidence formula (3.3) to arrive at an overall probability that s is the father. In particular, just as for identification, this probability involves a summation over all alternative fathers i, including the close relatives of s, and not just a single "random man" alternative father.

Weight-of-evidence for Forensic DNA Profiles   David Balding
© 2005 John Wiley & Sons, Ltd   ISBN: 0-470-86764-7

The weight-of-evidence formula is concerned with assessing the truth of the allegation, and not directly with any subsequent decisions. Thus, the assessment of evidence for paternity based on the formula is, in principle, the same whether the question arises as part of a criminal allegation against s or because of a civil dispute. In practice, these settings differ in that, roughly speaking, in criminal cases there is an *a priori* presumption of innocence in favour of the defendant, whereas in a civil case the court should treat the two sides equally. These differences can be incorporated into our Bayesian approach for evidence assessment in two ways:

- it is typically accepted that approximations should tend to favour s in the criminal setting (see Section 6.3), whereas this is often not appropriate in a civil case;

- a probability of paternity over 1/2 should suffice to prevail in a civil case in many legal systems, whereas for a criminal case a probability close to one is needed for a satisfactory conviction.

In other respects, the analyses should be the same in the two types of cases.

7.1.2 Prior probabilities

Despite the logical equivalence, DNA evidence is often assessed differently when paternity rather than identification is at issue. Many forensic scientists, lawyers, and academic commentators seem reluctant to consider in the paternity setting the very low prior probabilities that are now often accepted for identification. Indeed, there is a shamefully high prevalence of an unjustified assumption that both s and an unrelated "random man" i, have a prior probability 1/2 of being the father (Koehler 1993a). This "even prior" assumption is convenient since it implies that the likelihood ratio R_i is also the posterior probability that s is not the father. It may be based on the misguided assumption that equal prior weight should be assigned to the, typically conflicting, claims of s and m. However, this ignores the reality of, for example, additional background evidence and multiple alternative fathers.

Of course, with some exceptions such as paternity used to establish anonymous rape, paternity cases are in practice different from cases in which identification is a key issue because there is often substantial evidence in addition to the DNA evidence. Typically, s is known to m, they may have a blood relationship, and may have had an acknowledged sexual relationship. They may live near each other, possibly in the same dwelling. In such cases, there are few, if any, alternative fathers in a similar situation to s. Some commentators have argued that considering, say, a thousand alternative possible fathers is to slur m with the allegation that she had sexual intercourse with all of them. Others have tried to formulate prior probabilities on the basis of an estimate of the number of sexual partners of m. These lines of thought are both misguided: m may have had sex with one or a thousand men, the number is irrelevant. We know that she had sexual intercourse with at least one man, the father of c, and there may well be millions of possible

RELATEDNESS   113

candidates for consideration as being that one man, even if m had sex only once in her life.

The special status of s in a typical paternity dispute arises because of the substantial non-DNA evidence. In most settings, prior to assessing this evidence, s should be regarded as one of many candidates for being the father of c. However, since the non-DNA evidence is usually non-scientific and non-numerical, a court may wish to avoid formal application of the weight-of-evidence formula and informally arrive at assessments, based on the non-DNA evidence, for the probability of paternity of s and various alternatives. Alternatively, the court may wish to assess the DNA evidence using an equal prior probability for all sexually active men within some appropriate region, and assess the non-DNA evidence last (Section 9.3.4). In this case, an extremely small prior probability for s may well be appropriate. Attempting to assess a prior probability from the rate of exclusions in previous cases (Chakravarti and Li 1984) is misguided: background information relevant to the present case should be taken into account.

7.1.3 Calculating likelihood ratios

We will initially assume that i, m, and s have no direct relatedness. Direct relationships will be considered below, between i and s in Section 7.1.5, and between m and s (incest) in Section 7.1.6. The situation in which the profile of m is not available is discussed in Section 7.1.7. We will also (until Section 7.1.8) ignore the possibility of mutation. This means that R_i is undefined if m does not share an allele with c at every locus and is zero if the profile of s is inconsistent with he and m being the parents of c.

It is usually reasonable to assume that the probabilities that s and m have particular profiles is unaffected by whether s is the father of c, and it follows that we can rewrite (7.2) as a ratio of conditional probabilities for the child's profile:

$$R_i = \frac{P(\text{profile of } c \mid \text{profiles of } s \text{ and } m, \text{ father is } i)}{P(\text{profile of } c \mid \text{profiles of } s \text{ and } m, \text{ father is } s)}. \tag{7.3}$$

In essence, the difference between (7.2) and (7.3) is that in the latter the profile probabilities for s and m have cancelled, which can simplify subsequent formulas. However, if in similar problems such shortcuts cause any confusion or doubt, it is best to go back to the full likelihood ratio in which numerator and denominator both involve the probabilities of *all* the observed profiles.

Consider first a single locus, and assume that c's paternal allele can be determined (without assuming that s is the father), either because c is homozygous or because s/he shares exactly one allele in common with m. In this case, a man is excluded as a possible father if his profile does not include c's paternal allele. Then the probabilities for transmission of the maternal allele cancel in (7.3) to obtain

$$R_i = \frac{P(\text{paternal allele of } c \mid \text{profiles of } s \text{ and } m, \text{ father is } i)}{P(\text{paternal allele of } c \mid \text{profiles of } s \text{ and } m, \text{ father is } s)}. \tag{7.4}$$

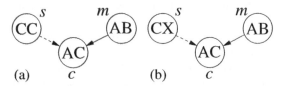

Figure 7.1 Single-locus genotypes in a case of questioned paternity in which the mother's genotype is available and suffices to determine the child's paternal allele (here, C). Ignoring mutation, a possible father must have genotype (a) CC or (b) CX, for some X $\neq$ C. The solid arrows indicate the known mother–child relationship, and the dashed arrows indicate putative father–child relationships.

For example, we may observe the genotypes of m and c shown in Figure 7.1, in which case we know that the child's paternal allele is C, and any man lacking a C allele is excluded from being the father of c.

Ignoring θ

The conventional approach to calculating likelihood ratios for paternity ignores coancestry, in effect assuming that $\theta = 0$. Under this approach, the profile of m plays no further role if it suffices to determine the child's paternal allele.

For the example of Figure 7.1, the denominator of (7.4) is 1 if s is homozygous for C and 1/2 if s has just one C allele. The numerator is the probability that an allele drawn from i is C. Since the profile of i is unavailable, this probability can be regarded as p_C, the proportion of C alleles in the population of i. Here, we will assume that p_C is known without error; in practice, an estimate is obtained from a population database (see Section 6.3.1). Then[1]

$$R_i = \begin{cases} p_C & \text{if } s \equiv \text{CC} \\ 2p_C & \text{if } s \equiv \text{CX} \end{cases} \qquad (7.5)$$

for any X $\neq$ C.

If $p_C > 0.5$ then $R_i > 1$ in the heterozygote case, and so the DNA evidence points away from s as a possible father, even though s has c's paternal allele. This situation rarely arises in practice but illustrates how likelihood ratios can sometimes clarify valid points that may seem counter-intuitive. Li and Chakravarti (1988) criticized the use of likelihood ratios for establishing paternity, arguing that it was "astonishing" that a hypothetical long sequence of such loci would suggest convincing evidence that s is not the father, even though he is not excluded at any locus from being the father. This may at first seem counter-intuitive, but some reflection shows that it is correct: see the discussion in Section 8.2.3.

---

[1]We noted in Section 3.2 that most authors report likelihood ratios as the inverse of mine: for example, here they would report $1/p_C$ in the homozygous case. In the paternity testing literature, this value is often referred to as a paternity index.

RELATEDNESS

Incorporating θ

The calculations leading to (7.5) ignore possible coancestry between i and one or both of s and m. If i is assumed to have coancestry with s, and possibly also m, measured by θ, then we can use the sampling formula (5.16) to compute the likelihood ratio (7.4).

Continuing with the example of Figure 7.1, the denominator of (7.4) is once again either 1 or 1/2, according to whether s is homozygous or heterozygous. In the numerator, we require the conditional probability of c's paternal allele given the profiles of s and m and the hypothesis that i is the father. Because of coancestry, the profiles of s and m may be informative about the allele transmitted by i to c, and so these profiles cannot automatically be neglected as was done when $\theta = 0$.

First, suppose that the coancestry of i and s is specified by θ but m has no coancestry with either (for example, she is from a different "racial" group). Then the profile of m is uninformative about the profile of i, and (7.4) becomes, in the homozygous and heterozygous cases, respectively,

$$R_i = \begin{cases} P(c \text{ has paternal allele C} \mid s \equiv CC, \text{ father is } i). \\ 2P(c \text{ has paternal allele C} \mid s \equiv CX, \text{ father is } i). \end{cases} \quad (7.6)$$

Since i is assumed to be the father, the profile of s is not directly relevant in (7.6), but it is indirectly relevant because i's profile is unknown and he may share alleles ibd with s due to coancestry.

In the homozygote case of (7.6), since we are given that i is the father, the paternal allele of c may be regarded as a third random draw from the population, and so the required probability is that a third allele is again C given that two observed alleles are both C. Invoking the recursive form of the sampling formula (5.6) with $m = n = 2$, we obtain

$$R_i = P(C \mid CC) = \frac{2\theta + (1-\theta)p_C}{1+\theta} \quad \text{if } s \equiv CC. \quad (7.7)$$

Reasoning similarly in the heterozygote case we have

$$R_i = 2P(C \mid CX) = 2\frac{\theta + (1-\theta)p_C}{1+\theta} \quad \text{if } s \equiv CX. \quad (7.8)$$

These expressions are reported in the final column of Table 7.1 (the heterozygote case is shown in rows 4, 7, and 9, according to the possible values for X, but R_i is the same in each case).

Suppose now that we assume that all three of i, m, and s have a common level of coancestry θ. Then we have

$$R_i = P(C \mid ABCC) = \frac{2\theta + (1-\theta)p_C}{1+3\theta} \quad \text{if } s \equiv CC$$

$$= 2P(C \mid ABCX) = 2\frac{\theta + (1-\theta)p_C}{1+3\theta} \quad \text{if } s \equiv CX. \quad (7.9)$$

Table 7.1 Single-locus likelihood ratios for paternity. The alleged father s is assumed unrelated to the alternative possible father, but they have nonzero coancestry, measured by θ; both are assumed here to have no coancestry with the mother m.

		Likelihood ratio $\times (1+\theta)$		
s	m	$c \equiv \text{AA}$	$c \equiv \text{AB}$	$c \equiv \text{AC}$
AA	AA	$2\theta + (1-\theta)p_A$		
AB	AA	$2(\theta + (1-\theta)p_A)$	$2(\theta + (1-\theta)p_B)$	
BB	AA		$2\theta + (1-\theta)p_B$	
BC	AA		$2(\theta + (1-\theta)p_B)$	$2(\theta + (1-\theta)p_C)$
AA	AB	$2\theta + (1-\theta)p_A$	$2\theta + (1-\theta)(p_A + p_B)$	
AB	AB	$2(\theta + (1-\theta)p_A)$	$2\theta + (1-\theta)(p_A + p_B)$	
AC	AB	$2(\theta + (1-\theta)p_A)$	$2(\theta + (1-\theta)(p_A + p_B))$	$2(\theta + (1-\theta)p_C)$
CC	AB			$2\theta + (1-\theta)p_C$
CD	AB			$2(\theta + (1-\theta)p_C)$

These, together with the formulae corresponding to other possible observed profiles, are given in Balding and Nichols (1995).

In both (7.8) and (7.9), we may have X = A or X = B; the result is unaffected (but recall that X ≠ C). Both heterozygote formulas reduce to $R_i = 2p_C$ if $\theta = 0$, in agreement with (7.5), and similarly the two formulas in the homozygous case reduce to $R_i = p_C$. The effect of possible coancestry between i and s is, because s carries a C allele, usually to make it slightly more likely that i also carries a C allele, which (slightly) lessens the evidential strength against s. This effect becomes more important as C becomes rarer. The direction of the effect can be reversed – so that consideration of coancestry can strengthen the evidence against s – in the heterozygote case if p_C is very large (larger than actually arises in practice).

Paternal allele ambiguous

Consider the case of Figure 7.2(a), for which the likelihood ratio is

$$R_i = \frac{P(c \equiv \text{AB} \mid m \equiv \text{AB}, s \equiv \text{AA}, \text{ father is } i)}{P(c \equiv \text{AB} \mid m \equiv \text{AB}, s \equiv \text{AA}, \text{ father is } s)}. \tag{7.10}$$

The denominator takes value 1/2 irrespective of any assumptions about θ, because there is probability 1/2 that m transmits allele B to c.

Ignoring θ, if i is the father then the profile of s is irrelevant, and we must consider the two possible transmissions of an allele from m to c, each of which has probability 1/2. In either case, c must have received the other allele from i, and hence the likelihood ratio is

$$R_i = \frac{(p_A + p_B)/2}{1/2} = p_A + p_B.$$

RELATEDNESS

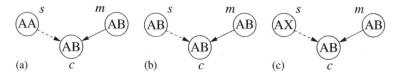

Figure 7.2 Single-locus genotypes in a case of questioned paternity in which the mother's genotype is available, but the child's paternal allele remains ambiguous. Three of the five possible paternal genotypes are shown, the remaining two are the same as (a) and (c) on replacing A with B. Here, $X \neq A$ and $X \neq B$.

If now we assume coancestry between i and s only, then R_i becomes

$$R_i = P(A \mid AA) + P(B \mid AA)$$
$$= \frac{2\theta + (1-\theta)p_A}{1+\theta} + \frac{(1-\theta)p_B}{1+\theta}$$
$$= \frac{2\theta + (1-\theta)(p_A + p_B)}{1+\theta}.$$

This formula is given in row 5, middle column, of Table 7.1.

For the case of Figure 7.2(b), the denominator of the likelihood ratio is 1/2, since there are two possible pairs of transmissions from m and s to c, and each possibility has probability 1/2. It then follows that R_i is the same as for Figure 7.2(a):

$$R_i = P(A \mid AB) + P(B \mid AB)$$
$$= \frac{\theta + (1-\theta)p_A}{1+\theta} + \frac{\theta + (1-\theta)p_B}{1+\theta}$$
$$= \frac{2\theta + (1-\theta)(p_A + p_B)}{1+\theta},$$

given in row 6, middle column, of Table 7.1. The likelihood ratio for the case of Figure 7.2(c) is given in row 7 of the table.

See Ayres and Balding (2005) for likelihood ratios given coancestry of all possible pairs from m, s, and i, Balding and Nichols (1995) for the case that all three of c, m, and s have coancestry at the same level, and Fung *et al.* (2004) for some examples of the effect of a θ-adjustment in some paternity cases.

7.1.4 Multiple loci: the effect of linkage

For loci on different chromosomes, the single-locus likelihood ratios considered above can be multiplied to obtain whole-profile likelihood ratios; the issues are the same here as discussed in Section 6.2.3. However, two loci in the CODIS system are both on chromosome 5, separated by approximately 25 centiMorgans. This implies that alleles at these loci are co-inherited in about 80% of parent–child transmissions, much greater than the 50% co-inheritance that applies to unlinked loci. This co-inheritance probability is far enough from 100% that, for unrelated individuals, it would not be expected to affect the validity of the "product rule" (taken here to mean multiplication across loci of appropriately adjusted match probabilities, see Section 6.2.3). However, for close relatives there is potentially a problem.

Nevertheless, linkage does not necessarily nullify the validity of the product rule. Suppose that s is heterozygous at two linked loci, A and B, having genotypes A_1A_2 and B_1B_2. If he transmits A_1 to his child, what is the probability that B_1 will also be transmitted? This scenario is illustrated in Figure 7.3(a). Because the loci are linked, whichever allele, B_1 or B_2, is in phase with A_1 is more likely to be transmitted with it. But this information is of no use if we are ignorant of the phase; in that case, the transmission probabilities are once again 50:50, just as in the unlinked case. This provides another reminder of the fact, discussed above in Section 6.2.3, that the independence of two events is not an absolute state of nature but a function of what we currently know (or assume).

Ignorance of phase only protects from the effects of linkage in simple pedigrees, such as a family trio of parents and child, in which each allele is involved in only one transmission. If genotypes are available on multiple transmissions within a pedigree, the co-inheritance pattern of alleles in one transmission is potentially informative about phase and hence affects the probabilities at the same locus in another transmission. For example, knowing that the man described above received A_1 and B_2 from his father makes it more likely that these two alleles are in phase and hence more likely that either both or neither will be transmitted to the man's child (Figure 7.3(b)). For a numerical example, see Buckleton *et al.* (2004).

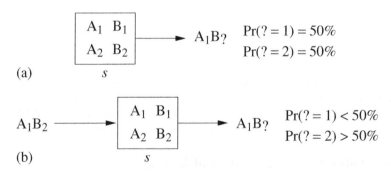

Figure 7.3 The effects of linkage on transmission probabilities. Inside the boxes are shown the genotypes of s at loci A and B. Arrows denote the parent – child transmission of a single haplotype. In (a), because phase is unknown, B_1 and B_2 are equally likely to be transmitted to the child. In (b), s received the A_1B_2 haplotype from his father, providing information on phase, and so B_2 is more likely than B_1 to be co-transmitted with A_1.

7.1.5 s may be related to c but is not the father

So far we have assumed no direct relationship between s and i, but in practice, we must very often assess the possibility that the man suspected of being the father is in fact a relative, such as a brother, of the true father i, and hence is related to c. In Table 6.3, we used $\bar{\kappa}$ to denote half the expected number of alleles at a locus shared ibd by i and s. Equivalently, $\bar{\kappa}$ is the probability that an allele chosen at random from i is ibd with one or other allele of s. Some values of $\bar{\kappa}$ for particular relationships between s and i were given in Table 6.3. (Note that these are twice Malécot's "coefficient of kinship", often denoted ϕ.)

Consider again the case that the paternal allele of c can be determined, say C, and that s is a CX heterozygote for some $X \neq C$. The denominator of the likelihood ratio (7.4) is once again 1/2. A child of i can receive a paternal C allele in two ways:

- a copy of allele C that is ibd with the C allele of s (probability $\bar{\kappa}/2$);
- an allele that is not ibd with either allele of s (probability $1 - \bar{\kappa}$) but happens to be allele C (probability p_C).

Thus, the likelihood ratio corresponding to (7.4) in the case that i and s are related is

$$R_i = \frac{\bar{\kappa}/2 + (1 - \bar{\kappa})p_C}{1/2} = \bar{\kappa} + (1 - \bar{\kappa})2p_C, \qquad (7.11)$$

which reduces to the heterozygote case of (7.5) when s and i are unrelated ($\bar{\kappa} = 0$). At the other extreme, if i and s are identical twins, $\bar{\kappa} = 1$ and (7.11) gives $R_i = 1$, corresponding to the fact that DNA evidence is of no value in distinguishing identical twins. For siblings, $\bar{\kappa} = 1/2$ and we obtain $R_i = 0.5 + p_C$. Similar to the unrelated case, discussed above in Section 7.1.3, if $p_C > 0.5$ then the observation that s is heterozygous favours his brother i as the father, rather than s himself.

Equation (7.11) has the form

$$R_i = \bar{\kappa} + (1 - \bar{\kappa})R_i^u, \qquad (7.12)$$

where R_i^u is the single-locus likelihood ratio that applies when i and s are unrelated. In fact, (7.12) is completely general, applying whatever be the genotypes of s, m, and c (see Balding and Nichols 1995). To understand why, consider the paternal allele of c under the hypothesis that i is the father:

- with probability $\bar{\kappa}$ it is ibd with an allele of s, in which case the situation is indistinguishable from the event that s is the father, and so the likelihood ratio is one;
- with probability $1 - \bar{\kappa}$ it is not ibd with an allele of s, in which case the situation is indistinguishable from the case that i and s are unrelated, for which the likelihood ratio is R_i^u.

In particular, (7.12) applies whether or not θ is taken into account, and so can be used in conjunction with Table 7.1, or the analogous expressions from Balding and Nichols (1995), to derive likelihood ratios that take into account both direct and indirect relatedness of i with s.

7.1.6 Incest

It may arise that s is acknowledged to be the father of m but is also accused of being the father of her child c (see Figure 7.4). Provided that the profiles of c, m, and s are consistent with the assertions that m is the mother of c, and s is the father of m, the effects of both these relationships cancel in the likelihood ratio, and they do not alter the resulting value. Thus, the likelihood ratio comparing the putative relationship shown in Figure 7.4 with the alternative that an unrelated man i is the father of c is the same as that calculated in a non-incest setting in Section 7.1.3.

However, if s is the father of both m and c, then with probability 1/2 he transmits the same allele to both of them and with probability 1/2 m transmits her maternal allele to c. Thus, with probability 1/4 c has both alleles ibd with his/her mother. An example of this is shown in Figure 7.4, where c has the same genotype as m which, if s is the father of both, has arisen because they both received the A allele from s.

In the heterozygote case, when m and c have the same genotype, the maternal and paternal alleles of c cannot be assigned, which typically leads to a larger likelihood ratio than when the paternal allele can be distinguished. Thus, when s is the father of c, the likelihood ratio will tend to be larger (less strong evidence) in incest cases than when s and m are unrelated. The same effect applies in the case of brother–sister incest, whereas for half-sib incest (as well as uncle/aunt–niece/nephew), the effect is half as strong: the probability of m and c having the same genotype ibd is 1/8 rather than 1/4.

Although the form of the likelihood ratio may be unaffected by incest, an appropriate prior probability may be especially difficult to specify; see the discussion of Section 7.1.2. It may also be important to consider other male relatives of m in addition to s.

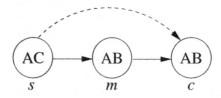

Figure 7.4 Relationships in an incest case. Circles represent individuals, solid arrows indicate accepted parent–child relationships, and the dashed arrow indicates the putative relationship in question. Single-locus genotypes consistent with the putative relationship are also shown.

7.1.7 Mother unavailable

If the profile of m is unavailable, the appropriate version of the likelihood ratio is

$$R_i = \frac{P(\text{profile of } c \mid \text{profile of } s, \text{ father is } i)}{P(\text{profile of } c \mid \text{profile of } s, \text{ father is } s)}. \quad (7.13)$$

For example, we may require

$$R_i = \frac{P(c \equiv AA \mid s \equiv AA, \text{ father is } i)}{P(c \equiv AA \mid s \equiv AA, \text{ father is } s)}.$$

Allowing for coancestry between s and i, the denominator is

$$P(c \equiv AA \mid s \equiv AA, \text{ father is } s) = \frac{P(AAA)}{P(AA)} = \frac{2\theta + (1-\theta)p_A}{1+\theta},$$

while the numerator is

$$P(c \equiv AA \mid s \equiv AA, \text{ father is } i) = \frac{P(AAAA)}{P(AA)}$$

$$= \frac{P(AAA)}{P(AA)} \times \frac{3\theta + (1-\theta)p_A}{1+2\theta},$$

and so

$$R_i = \frac{3\theta + (1-\theta)p_A}{1+2\theta}.$$

Likelihood ratios for this and other cases have been given by Ayres (2000b) and are shown in Table 7.2. Ayres and Balding (2005) give likelihood ratios for the coancestry of all possible pairs of m, s, and i in the m-unavailable case (note that coancestry of s or i with m can have an effect, even if m is not typed, since one of her alleles at each locus is observed in c.).

Table 7.2 Single-locus likelihood ratios (R_i) for paternity when the mother's profile is unavailable. The alleged father s is assumed unrelated to the alternative possible father, but they have coancestry measured by θ. These formulas were first reported (as inverses of those given here) in Table 1 of Ayres (2000b).

c	s	$R_i \times (1+2\theta)$
AA	AA	$3\theta + (1-\theta)p_A$
AA	AB	$2(2\theta + (1-\theta)p_A)$
AB	AA	$2(2\theta + (1-\theta)p_A)$
AB	AC	$4(\theta + (1-\theta)p_A)$
AB	AB	$4(\theta + (1-\theta)p_A)(\theta + (1-\theta)p_B)/(2\theta + (1-\theta)(p_A + p_B))$

7.1.8 Mutation

Mutation rates for STR loci are about 1 or 2 per 1 000 generations, and so if an STR profile consists of around ten loci, there is very roughly a 2% probability that there will be a mutation in transmission from father to child, in which case a strict use of likelihood ratios that ignore possible mutations would lead to a false exclusion.

The question arises as to whether, given profiles that are consistent with s being the father of c at many loci, but an apparent exclusion at just one or perhaps two loci, it might be possible to report strong evidence that s is the father, and that one or more mutations have occurred. To answer this question quantitatively, we need to compute the appropriate likelihood ratios, of the form (7.3) but now allowing for mutation. Dawid *et al.* (1996) give a systematic treatment of this question when the mother's profile is available, but ignoring coancestry. The latter omission is rectified by Ayres (2002), who treats the case that any two of m, s, and i have the same level of coancestry, θ, correcting formulas given in Ayres (2000b) that neglected the possibility of mutations occurring during maternal transmission. Here, we work through three particular examples, shown in Table 7.3, and briefly discuss some additional issues not considered by these authors.

First, consider a general mutation model in which any allele can be transformed into any other allele in transmission from parent to child. We write C_1 and C_2 for the child's two alleles, ordered arbitrarily, while the mother's and the alleged father's two alleles are labelled M_1, M_2, S_1, and S_2. Since any mutation is possible, under the hypothesis that s is the father C_1 could have descended from any one of the four parental alleles, and for each of these, there are two possibilities for the parental allele from which C_2 has descended. Four of these eight possibilities for the pair of transmissions are shown in Figure 7.5(a). Thus, there are in principle eight terms contributing to the denominator of (7.3). However, since mutations are rare, any term corresponding to one (respectively, two) mutations can, in practice, be neglected if there is also at least one term corresponding to zero (respectively, one) mutation.

Since the alternative father i is unprofiled, under the hypothesis that he is the father, the allele immediately ancestral to c's paternal allele is unobserved. Thus,

Table 7.3 Three examples of scenarios in which a mutation is required to sustain the hypothesis that m and s are the parents of c.

Scenario		(i)	(ii)	(iii)
Child c	≡	AB	AB	AB
Mother m	≡	AC	AC	DD
Alleged father s	≡	CD	AD	AC

RELATEDNESS

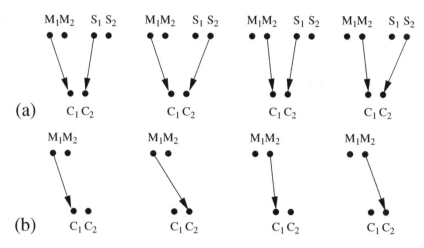

Figure 7.5 (a) Four of the eight possible transmissions of alleles from m and s to c. The other four are obtained by interchanging C_1 with C_2. (b) The four possibilities for the transmission of an allele from m to c.

mutation can always be ignored in the paternal transmission under this hypothesis, and the numerator of (7.3) has the four terms corresponding to the possible maternal transmissions indicated in Figure 7.5(b). If c has at least one allele in common with m, then terms involving a mutation can be neglected.

Table 7.3, case (i)

Except for relabelling of alleles, this case is the same as that shown in row 15 of Table 1 of Dawid *et al.* (1996). The genotypes are consistent with c having received A from m, but c has no allele in common with s. The transmissions consistent with the minimal number of mutations under the hypotheses that (a) s is the father, and (b) an unprofiled man i is the father, are shown in Figure 7.6. A

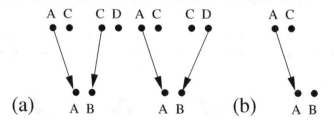

Figure 7.6 Allele transmissions for case (i) of Table 7.3. (a) Under the hypothesis that s is the father, the two possible transmissions of alleles from m and s to c consistent with one mutation. (b) The one possible transmission of alleles from m to c consistent with no mutation.

factor of 1/2 for the transmission from m to c of allele A cancels in both numerator and denominator of (7.3). Ignoring this term, and θ, the numerator is p_B while the denominator is $(\mu^f_{C \to B} + \mu^f_{D \to B})/2$, in which we introduce $\mu^f_{X \to Y}$ to denote the probability that allele X mutates to allele Y in transmission from father to child. Thus, the likelihood ratio is

$$R_i = \frac{P(\text{paternal allele of } c \text{ is B} \mid s \equiv \text{CD, father is } i)}{P(\text{paternal allele of } c \text{ is B} \mid s \equiv \text{CD, father is } s)}$$

$$= \frac{2p_B}{\mu^f_{C \to B} + \mu^f_{D \to B}}$$

Assuming now that i and s have coancestry at level θ, but each has no coancestry with m, the probability that an allele drawn from i is B, given that $s \equiv$ CD, is $(1 - \theta)p_B/(1 + \theta)$. Thus, the likelihood ratio is

$$R_i = \frac{2(1 - \theta)p_B}{(1 + \theta)(\mu^f_{C \to B} + \mu^f_{D \to B})}.$$

The case in which all three of m, s, and i have coancestry is given in Ayres (2002): the $1 + \theta$ in the denominator is replaced by $1 + 3\theta$. Under either assumption, since only one B allele is observed, the effect of a θ adjustment is to (slightly) reduce R_i, strengthening the case that s is the father.

Table 7.3, case (ii)

This case is equivalent to that of row 13 of Table 1 of both Dawid *et al.* (1996) and Ayres (2002). Both m and s share an A allele with c but neither shares c's B allele. Thus a mutation must have occurred in transmission either from m or from s, if these are the parents of c. The four possibilities for transmissions involving one mutation are shown in Figure 7.7. The event that m transmits allele A to c with no mutation and s transmits either of his alleles and it mutates to B has probability $(\mu^f_{A \to B} + \mu^f_{D \to B})/4$, while the event that s transmits allele A to c with no mutation and m transmits either of her alleles and it mutates to B has probability $(\mu^m_{A \to B} + \mu^m_{C \to B})/4$, where the superscript m denotes that the mutation rate is for

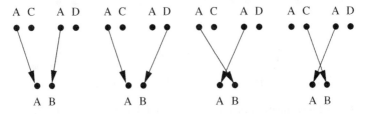

Figure 7.7 Allele transmissions for case (ii) of Table 7.3 under the hypothesis that s is the father. The four possible transmissions of alleles from m and s to c consistent with one mutation are shown.

RELATEDNESS

maternal transmission (typically lower than the rate for paternal transmissions). The numerator of (7.3) is $p_B/2$, and so

$$R_i = \frac{2p_B}{\mu^m_{A \to B} + \mu^m_{C \to B} + \mu^f_{A \to B} + \mu^f_{D \to B}}.$$

The effect of a θ adjustment is to reduce R_i by the same factor as in case (i), that is either $(1-\theta)/(1+\theta)$ if just the coancestry of s and i is considered, or $(1-\theta)/(1+3\theta)$ if m also has coancestry with s and i.

Table 7.3, case (iii)

Here, s shares an A allele with c, but m has no allele in common with c and so is excluded as the mother unless a mutation has occurred. If such a pattern arises at two or more loci, we should seriously question the proposed maternal relationship. However, if this relationship is accepted, then at least one mutation must have occurred, and as usual, we neglect the possibility of more than one mutation. Thus, one of the D alleles of m must have mutated to B in transmission to c; the two possibilities are indicated in Figure 7.8. These possibilities are the same irrespective of whether s or i is the father, and so the terms corresponding to them cancel in the likelihood ratio, which is the same as in the no mutation case when the profile of m is unavailable, given in row 4 of Table 7.2. Thus, the non-match of an allele between m and c has essentially no effect on the likelihood ratio, except that it implies that the paternal allele of c is ambiguous.

Mutation models

We noted in Section 5.1.2 that not all STR mutations are equally likely. Single-step mutations, in which allele length changes by one repeat unit, are more frequent than multi-step mutations. A simple model that approximates this reality is the SMM, in which both single-step mutations are assigned probability μ whereas multi-step mutations have probability zero. Thus, under the SMM, s is regarded as excluded from being the father if at any locus a multi-step mutation is required to sustain that hypothesis.

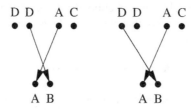

Figure 7.8 Allele transmissions for case (iii) of Table 7.3 under the hypothesis that s is the father. The two possible transmissions of alleles from m and s to c consistent with one mutation are shown.

Table 7.4 Single-locus likelihood ratios for paternity when the mother's profile is unavailable and when a mutation must have occurred if s is the father of c. The alleged father s is assumed unrelated to the alternative possible father, but they have coancestry, measured by θ. These formulas are reported as inverses in Table 4 of Ayres (2000b), but the mutation rates M used there are double the μ used here.

c	s	$R_i \times (1 + 2\theta)$
AA	BB	$(\theta + (1-\theta)p_A)/\mu^f_{B \to A}$
AA	BC	$2(\theta + (1-\theta)p_A)/(\mu^f_{B \to A} + \mu^f_{C \to A})$
AB	CC	$2(1-\theta)p_A p_B/(\mu^f_{C \to A} + \mu^f_{C \to B})$
AB	CD	$4(1-\theta)p_A p_B/(p_A(\mu^f_{C \to B} + \mu^f_{D \to B}) + p_B(\mu^f_{C \to A} + \mu^f_{D \to A}))$

Dawid *et al.* (1996) discuss the SMM and more general mutation models. These authors give an example of 13-locus profiles of m, s, and c indicating exclusions of s as the father at two loci. They calculate an overall likelihood ratio (not allowing for coancestry) that indicates weak evidence for the non-paternity of s. However, if one of the loci showing exclusions had been omitted, the likelihood ratios would indicate moderately strong evidence for paternity, despite the one remaining exclusion.

Mother unavailable

Likelihood ratios in the case of apparent exclusion of the alleged father s when the mother's genotype is unavailable are shown in Table 7.4. The final two values differ by a factor of two from those reported by Ayres (2000b) because of a different definition of mutation rates.

7.2 Other relatedness between two individuals

7.2.1 Only the two individuals profiled

If the two individuals are genotyped at a single-locus, and assuming that no profiles of their relatives are available, the likelihood ratio comparing a specified regular relationship with the hypothesis that the individuals are unrelated is (here, small values support no relationship):

$$R = \kappa_0 + \kappa_1/R_i^p + \kappa_2/R_i^u, \qquad (7.14)$$

in which the κ_j are the relatedness coefficients given in Table 6.3, and

- R_i^p denotes the single-locus likelihood ratio for paternity in the unrelated case when the mother's genotype is unavailable; values are given in Table 7.2 when no mutations are required to sustain the parent–child hypothesis, and in Table 7.4 when mutations are required (in practice, we can set $1/R_i^p = 0$ in these cases if $\kappa_1 < 1$).

- R_i^u denotes the single-locus likelihood ratio for identity in the unrelated case. When the genotypes of the two individuals match, this is the match probability given at (6.3) and (6.4). Otherwise we interpret $1/R_i^u$ as zero, in effect ignoring mutation, which would be important only in the case of identical twins, $\kappa_2 = 1$.

Explicit formulas corresponding to (7.14), and neglecting mutations, when

- $\kappa_1 = 1/2$, $\kappa_2 = 1/4$, (sibling);
- $\kappa_1 = 1/2$, $\kappa_2 = 0$, (half-sib, uncle/nephew, grandparent/grandchild);
- $\kappa_1 = 1/4$, $\kappa_2 = 0$, (cousin),

are given in Table 3 of Ayres (2000b). Fung et al. (2004) also give general expressions and explicit examples. Brenner and Weir (2003) describe particular problems arising in the application of STR profiles to identify victims of the New York World Trade Center disaster in 2001.

7.2.2 Profiled individual close relative of target

The first prosecution arising from identification of a relative through a partial DNA profile match occurred in Surrey, UK, in April 2004. The convicted man had no criminal record and no match of the crime-scene profile was found in the UK national DNA database, which contains the profiles of over 10% of the adult male population. However, a partial match, of 16 alleles out of 20, was found, and this led police to investigate the immediate family of the individual providing the partial match. A full DNA profile match was obtained with a brother, who was charged and eventually convicted.

The possibility that an observed DNA profile, partially matching a crime-scene profile, comes from a close relative of the target individual can be assessed quantitatively using likelihood ratios (7.14). Since the general population contains very many people unrelated to a given individual, there is potentially a problem that, by chance, some unrelated individuals resemble the crime-scene DNA profile as closely as would be expected from a true relative. This is known as the "multiple testing" problem in classical statistics, but this is a misnomer since the problem is not the multiplicity of the tests (i.e. the number of individuals searched in the database), rather it is that, in a national investigation, most pairs of individuals are unrelated. Thus, close relatives are *a priori* unlikely, and so substantial evidence is needed to be convinced of a brother relationship. Appropriate prior probabilities can be based on average population values. For example, in a population of 5 million adult males in which the average number of brothers of any man is 0·5, the prior probability that two given men are brothers could be taken to be 1 in 10 million.

The operating characteristics of procedures for deciding whether the crime-scene profile comes from a brother of an individual whose DNA profile is available were investigated by Sjerps and Kloosterman (1999). On the basis of a particular eight-locus STR profile, they estimated that around 5% of pairs of unrelated individuals in the Dutch population would have $R > 1$ (falsely suggesting fraternity), and about the same proportion of pairs of brothers would have $R < 1$, wrongly suggesting unrelatedness. Conversely, they found that nearly 80% of unrelated pairs of individuals had $R < 0.1$, and nearly 60% of brothers had $R > 100$. This suggests that the 10 to 15 loci used in forensic work will often be very helpful in suggesting a brother relationship, but will rarely suffice for a definitive establishment of such a relationship, particularly when other possible relationships, such as half-sibling or uncle–nephew, must be taken into account. Rather than only contrast the sibling relationship with unrelated, a more satisfactory approach is to use (7.14) to calculate a probability distribution for all the possible relationships of two profiled individuals (see Egeland et al. 2000).

7.2.3 Profiles of known relatives also available †

When a putative relationship between two individuals is being investigated via their DNA profiles, it can be helpful, when available, to have the profiles of known relatives of one or both of the individuals. For example, if the putative relationship is half-sibling through a common father, then knowing the profiles of the two mothers (here assumed unrelated) can strengthen the evidence for or against a common father. Perhaps surprisingly, when coancestry can be neglected (i.e. $\theta = 0$), there are only four distinct forms for the LR:

(i) It can be verified that the children do not share a paternal allele, for example, child 1 $\equiv$ AB, child 2 $\equiv$ CD. In these cases the LR is 1/2.

(ii) Both paternal alleles can be identified and are the same, or the paternal allele of one child can be identified and is the same as one of the other child's alleles. An example of such genotypes, and the corresponding likelihood ratio, is shown in row (ii) of Table 7.5.

(iii) Neither paternal allele can be identified, and the children have the same (heterozygous) genotypes. See row (iii) of Table 7.5.

(iv) Neither paternal allele can be identified, and the children are both heterozygous but share only one allele. See row (iv) of Table 7.5.

For highly polymorphic loci, case (i) applies frequently when the children are unrelated, whereas without the mothers' genotypes a likelihood ratio of 1/2 is much less frequent. The 10 to 15 STR loci currently in routine forensic will rarely suffice to convincingly establish a relationship as distant as half-sibling (misclassification rates typically 5% to 10%), but the inclusion of the mothers' profiles can strengthen the evidence considerably and, surprisingly, is on average slightly more informative

Table 7.5 Examples of single-locus likelihood ratios, ignoring θ, comparing the hypothesis that two children have a common father with that of unrelated fathers, with and without the genotypes of the children's mothers (assumed unrelated). Large values of the likelihood ratio support the half-sibling relationship.

	Genotypes		Likelihood ratio	
	Mother	Child	With mothers	Ignoring mothers
(i)	AB AB	AC AD	$\frac{1}{2}$	$\frac{1}{2}\left(1+\frac{1}{4p_A}\right)$
(ii)	AC AC	AB AB	$\frac{1}{2}\left(1+\frac{1}{p_B}\right)$	$\frac{1}{2}\left(1+\frac{1}{4p_A}+\frac{1}{4p_B}\right)$
(iii)	AB AB	AB AB	$\frac{1}{2}\left(1+\frac{1}{p_A+p_B}\right)$	$\frac{1}{2}\left(1+\frac{1}{4p_A}+\frac{1}{4p_B}\right)$
(iv)	AB AC	AB AC	$\frac{1}{2}\left(1+\frac{p_A}{(p_A+p_B)(p_A+p_C)}\right)$	$\frac{1}{2}\left(1+\frac{1}{4p_A}\right)$

than typing the same number of additional loci in the putative half-sibs (Mayor and Balding 2005).

7.3 Software for relatedness analyses

`Familias` computes likelihoods and hence posterior probabilities for arbitrary pedigrees, given DNA data from some of the individuals. It permits various DNA data types and mutation models, incorporates a θ adjustment for coancestry, and is freely available for Microsoft platforms, together with documentation, from http://www.nr.no/familias. Its use is also described in Egeland *et al.* (2000), who discuss selecting the most probable pedigrees to consider, when there are too many possible pedigrees to compute likelihoods for. Another set of programs for paternity analysis is described in Fung (2003).

Bayesian networks (BN), also known as probabilistic expert systems, form a flexible class of computer software that implements Bayesian statistical models in such a way that the probability distributions of unknowns can be efficiently computed (via Bayes Theorem). Moreover, the probability distributions are readily updated when additional information is incorporated. The statistical model underlying a BN can be represented graphically: data and variables are represented by nodes, and these may be connected by directed arcs (arrows), signifying dependence relationships. Intuitively, an arc from node B to node C indicates that information about the value of B is potentially also informative about the value of C. Informally, C is said to be the "child" of B, and conversely B is the "parent" of

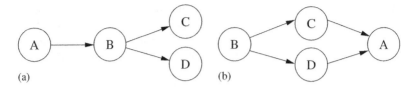

Figure 7.9 Two directed, acyclic graphs, each with four nodes.

C. Figure 7.9 illustrates two directed graphs in which B is the parent of both C and D.

Given the values of all the parents of a given node in the graphical representation of a BN, the value at that node is independent of the values at all nodes other than its descendants. For the graphs of Figure 7.9, D is conditionally independent of C given B; in graph (a), D is also conditionally independent of A given B, but this is not so for graph (b). This independence is *conditional* and any two of the variables are dependent unconditionally. This conditional independence structure is the key to the computational efficiency of BN for many complex problems, since it allows them to be broken down into components. The definition of BN requires the underlying directed graph to be acyclic, which means that it has no paths (followed in the direction of the arcs) that end where they started. Equivalently, no node can be its own descendant. The "loop" of Figure 7.9(b) is allowed because of the direction of the arcs (for example, there is no arc away from A).

BN are potentially very useful for analysing problems of unknown relationships. Their flexibility permits, in principle, multiple individuals with some relationships known, additional background information, missing data, and mutation. The language of "parent" and "child" suggests that the pedigree could serve as the directed, acyclic graphs on which an appropriate BN could be based. This would be the case for haploid species. For diploids, although a valid BN can be specified at the level of a pedigree, the need to separately track the inheritance of an individual's two alleles at an autosomal locus means that it is typically more useful to specify a richer graphical structure in which nodes correspond to alleles rather than individuals (see Lauritzen and Sheehan 2003 for a further discussion and an excellent, brief introduction to BN applied in genetics). As for all statistical modelling, there is no one "correct" BN for the analysis of a relatedness problem: different representations of the problem can have different advantages, or reflect different assumptions.

Dawid *et al.* (2002) give several examples of BN specified for the assessment of different forensic problems involving relatedness. They do not consider the incorporation of a θ-correction to allow for coancestry. This would add considerably to the complexity of the BN because the alleles for which no parent is included in the model would no longer be independent.

Beyond relatedness problems, BN are potentially useful for many forensic applications involving a wide range of evidence types. For a gentle introduction in the context of general forensic applications, see Aitken and Taroni (2004). Mortera *et al.* (2003) describe applications of BN to the analysis of DNA mixtures (Section 6.5).

It is possible that BN will eventually revolutionize the interpretation of forensic evidence. For the moment, there seems to be no BN software widely available that is tailored for forensic applications. Forensic problems can be tackled using currently available general-purpose BN software, but this is often far from straightforward. For problems involving the assessment of relatedness via DNA evidence, Cowell (2003) described prototype BN software that, at the time of writing, seems not to have been released.

7.4 Inference of ethnicity or phenotype †

When a DNA profile is obtained from a crime scene, it may be of assistance to investigators to predict the ethnic origin of its unknown source. Such predictions are complicated by the fact that "ethnicity" and "race" largely reflect social context rather than genetics, and "ethnic" groups are not precisely defined. In particular there are different possible levels of classification; for example "African", "Nigerian", and "Ibo" may all be applied to the same individual. Although different human groups have obvious phenotypic differences, these correspond to only a small part of the genome, and for typical loci humans form a remarkably homogeneous species at the DNA level. We are close to being the most homogeneous of large-bodied mammals, despite our worldwide distribution (Templeton 1999). Moreover, human allele frequencies usually vary smoothly with distance, with few sharp changes caused by geographic or social barriers. In a classic study, Lewontin (1972) found that about 85% of human protein variation was between individuals within population groups, while the remaining 15% was roughly equally divided into variation between populations within continental groups, and variation between continents. More recent studies involving much larger datasets of DNA markers have shown broadly the same pattern (Jobling *et al.* 2004, Table 9.1).

For these reasons, and perhaps also because of sensitivity to social and political concerns, many authors have argued that notions of "ethnicity" and "race" have no useful role in genetics-related science. An alternative view is that in most nations there are recognizable, distinct groups into which most of the population can reliably be allocated, for example, by self-report. These groups often have enough genetic differences that classification of a query sample of DNA into its source group can be performed with reasonable accuracy: sufficient to be of some use to crime investigators, provided that the limitations of the approach are appreciated.

Suppose that the population can be classified into K groups. Then, by an application of Bayes Theorem (Section 3.5), the posterior probability that a DNA sample with profile D comes from group k can be written as

$$P(\text{group } k \mid D) = \frac{P(D \mid \text{group } k)\pi_k}{\sum_{j=1}^{K} P(D \mid \text{group } j)\pi_j}, \qquad (7.15)$$

in which π_j denotes the prior probability of group j. Population proportions of the different groups may provide acceptable values for the π_j, or it may be possible to use more specific local information, or possibly other information on the different patterns of offending recorded for the different groups.

If we assume Hardy–Weinberg equilibrium (Section 5.4) and linkage equilibrium (Section 5.5), then each likelihood $P(D \mid \text{group } j)$ can be expressed in terms of a product over allele proportions in group j (Rannala and Mountain 1997). Of course, true allele proportions in the different groups will not be known, but these can be estimated from samples. A problem arises for rare alleles with a low sample count, since small differences attributable to stochastic variation could then lead to very big differences in the likelihood used. Rannala and Mountain (1997) in effect employed a "pseudo-count" approach to overcoming this problem, in which each sample count was incremented by the inverse of the number of distinct alleles at the locus. We have advocated a similar approach above in Section 6.3.1, except that we have employed a pseudo-count of one. Lowe *et al.* (2001) employed a minimum allele proportion to overcome the rare-allele problem.

Provided that the rare-allele problem is overcome, the above approach can give some information about the ethnic origin of an unknown offender. For example, a forensic scientist may be able to advise the police that *provided that* the offender (who left a DNA stain at the crime scene) comes from one of the major population groups (such as "European", "Asian", and "African"), s/he evaluates probabilities of 88% that the offender is of European origin, 9% that the Asian origin, and 3% offender is of African origin. Because of the limited discriminatory power of forensic STR loci, it will be unlikely that a small minority group will be correctly identified as the source population. Also, no assessment can easily be made of the probability that the offender comes from outside those groups for whom allele-frequency estimates are available. Similarly, assignment of individuals with mixed or otherwise unusual ancestry can be difficult. But, provided that these limitations are understood, there is the potential for crime investigations to be assisted.

Assessing the effectiveness of a classification technique can be challenging. There are several pitfalls that can lead to exaggerated claims of accuracy, for example a bias arising through testing a classifier with the same data as were used to fit the model. See, for example, the criticism by Brenner (1998) of the analyses of Shriver *et al.* (1997), as well as the reply (Shriver *et al.* 1998).

Romualdi *et al.* (2002) found that 21 "random" diallelic autosomal markers did not suffice to achieve a probability above 80% of correct classification according to continent of origin, but Bamshad *et al.* (2003) obtained near 100% success in discriminating between three source continents with 100 such loci. Rosenberg *et al.* (2003) have identified STR markers with good discriminatory power for identifying geographical origin, but the STR markers in routine forensic use tend to have been selected to have relatively uniform proportions across populations. This is advantageous for calculating match probabilities, but disadvantageous for inferring ethnic origin. In fact, Lowe *et al.* (2001) reported a success rate of over 80% using 6 forensic STR loci when only two populations were considered, each equally likely *a priori* to be the source. Success rates were only around 50% for five equally likely source populations. Recently, Frudakis *et al.* (2003) have reported very high predictive accuracy for SNP-based ethnic classification.

Another approach is to predict some aspects of the offender's phenotype from a crime-scene sample. Currently, the offender's sex is reliably predicted via the Amelogenin locus (page 46). In the near future, it seems likely that from a query DNA sample it will be possible to predict (imprecisely) the height, weight, skin and hair colour, and some other characteristics of an offender. Currently, however, although many alleles have been characterized that have large phenotypic effects, these are mostly rare and so are of little forensic use. Very little of ordinary phenotypic variation has as yet been characterized; the exceptions include some genes involved in determining skin and hair colour, some genetic variants for the digestion of alcohol and lactose, and some alleles contributing to obesity. Of these, only the genetic polymorphisms underlying the phenotype "red hair" are currently well characterized, although further rapid developments are expected. See Jobling *et al.* (2004) for further details and references.

7.5 Relatedness exercises

1. Consider three-locus STR profiles as follows:

	Locus		
Individual	1	2	3
Child c	AB	AA	AB
Mother m	AC	AA	AB
Alleged father s	BC	AB	AC

   Find the overall likelihood ratio comparing the hypothesis that s is the father of c with the hypothesis that a man i, unrelated to both s and m, is the father:

   (a) assuming no coancestry (i.e. $\theta = 0$);

   (b) assuming $\theta = 5\%$ for s and i, but no direct relatedness of these two with m.

   (c) †assuming $\theta = 5\%$ for each of s, m, and i.

   At each locus, assume that $p_A = 10\%$ and $p_B = 5\%$.

2. (a) Repeat 1(a) but now assume that i is the half-brother of s.

   (b) What is the posterior probability that s is the father of c if there are 12 potential fathers: s, his half-brother, and 10 unrelated men. Use $\theta = 0$, and assume that each man is equally likely *a priori* to be the father.

3. Repeat 1(a) but now assume that genotypes from a fourth STR locus are also available. At this locus $c \equiv AA$, $m \equiv AB$, and $s \equiv CC$. Here, assume that every C allele has probability 1 in 1/2000 of mutating to an A in each paternal transmission. (Once again assume that $p_A = 10\%$ and $p_B = 5\%$).

4. Using only the three-locus profiles of s and c given in 1 above, calculate an overall likelihood ratio comparing

    (a) the hypothesis that they are siblings with the alternative that they are unrelated; use (i) $\theta = 0$, and (ii) $\theta = 2\%$.

    (b) †The hypothesis that they are siblings with the alternative that they are father–child; use $\theta = 0$.

8

Other approaches to weight of evidence

We have seen that the approach to assessing evidence via likelihood ratios and Bayes Theorem is very powerful. For DNA evidence, no matter what unusual circumstances arise in a new case: identical twins, inbreeding, fraud or missing bands, the likelihood ratio provides us with a framework for assessing the evidence. For every piece of evidence, the finder of fact (e.g. a juror) should ask two questions:

- How likely is the evidence if the defendant s is guilty?

- How likely is the evidence if s is innocent and i is the true culprit?

Consequently, expert witnesses should provide as much information as possible to help jurors to answer these questions.

Although elegant and powerful, the weight-of-evidence theory based on likelihood ratios is often viewed as complicated and unfamiliar. The real world, and actual crime cases in it, are themselves complicated, so to some extent it is inevitable that a satisfactory theory of evidential weight cannot be too simple. Here, we briefly introduce alternative approaches that seem simpler. Although not without advantages, each of these approaches has hidden difficulties.

8.1 Uniqueness

Match probabilities for alternative possible culprits unrelated to the defendant are often extremely small: for STR systems with 10 or more loci that are currently in widespread use, calculated match probabilities are usually substantially less than 1 in 1 billion. When a forensic scientist reports match probabilities this small,

Weight-of-evidence for Forensic DNA Profiles David Balding
© 2005 John Wiley & Sons, Ltd ISBN: 0-470-86764-7

it seems effectively equivalent to saying that s/he is reasonably certain that the defendant's DNA profile is unique in the population of possible sources of the crime stain. If so, would not jurors be better assisted by the expert giving a "plain English" statement of this, rather than a match probability whose unfamiliar magnitude may overwhelm or confuse? For example, perhaps an expert witness could assert that, excluding identical twins and laboratory/handling errors, in his/her opinion the defendant's DNA profile is almost certainly unique in some appropriate population.

Although attractive in some respects, a practice of declaring uniqueness in court does lead to difficulties. One of these is how to deal with the minority of cases in which uniqueness cannot reasonably be asserted? These may arise for mixed profiles, or in cases of questioned paternity or other relatedness (Chapter 7). Perhaps the most important barrier to declaring uniqueness is the problem of the non-DNA evidence in a case. The event that a particular DNA profile is unique is either true or false and no "objective" probability can be assigned to it. Nevertheless, since this truth or falsity cannot be established in practice, a probability of uniqueness based on the information available to an expert witness, such as that obtained from population databases of DNA profiles, together with population-genetics theory, may potentially be useful to a court. The problem then arises as to what data and theory the expert should take into account. Specifically, the non-DNA evidence in a case may be directly relevant, yet it may not be appropriate for the DNA expert to assess this evidence.

Consider a crime-scene DNA profile that is thought to be so rare that an expert might be prepared to assert that it is unique. Suppose that, for reasons unrelated to the crime, it is subsequently noticed that the crime-scene profile matches that of the Archbishop. On further investigation, it is found to be a matter of public record that the Archbishop was taking tea with the Queen at the time of the offence in another part of the country. A reasonable expert would, in the light of these facts, revise downwards any previous assessment of the probability that the crime-scene profile was unique. However, this is just an extreme case of the more general phenomenon that any evidence in favour of a defendant's claim that he is not the source of the crime stain is evidence against the uniqueness of his DNA profile.

8.1.1 Analysis

Let U denote the event that both

- the DNA profile of s, the defendant, matches the crime-scene profile; and
- there is no matching individual in $\mathcal{P}$, the population of alternative possible culprits.

It is also assumed that s is known to have no identical twin, that laboratory or handling errors do not occur, and that the crime-scene profile is that of the true culprit. Under these assumptions, U implies G, and so

$$P(U \mid E) \leq P(G \mid E),$$

OTHER APPROACHES TO WEIGHT OF EVIDENCE

where E here denotes the DNA evidence; other background evidence and information is suppressed in the notation. Using (3.3) we have

$$P(U \mid E) = P(U \mid G, E)P(G \mid E) > \frac{P(U \mid G, E)}{1 + \sum_{i \in \mathcal{P}} w_i R_i}, \qquad (8.1)$$

where R_i denotes the match probability for alternative possible culprit i, while w_i measures the weight of the non-DNA evidence against i, relative to its weight against s.

It seems reasonable to suppose that the DNA evidence in the case has no bearing on the question of uniqueness beyond its bearing on the question of guilt, so that $P(U \mid G, E) = P(U \mid G)$. Further, non-uniqueness is the union of the events "the DNA profile of i matches the DNA profile of s", for all $i \in \mathcal{P}$. From the elementary rules of probability, the probability of this union is bounded above by the sum of all the match probabilities (which equals the expected number of individuals in $\mathcal{P}$ sharing the DNA profile of s). It follows that

$$P(U \mid G, E) = P(U \mid G) > 1 - \sum_{i \in \mathcal{P}} R_i. \qquad (8.2)$$

Substituting (8.2) in (8.1) gives

$$P(U \mid E) > \frac{1 - \sum_{i \in \mathcal{P}} R_i}{1 + \sum_{i \in \mathcal{P}} w_i R_i}. \qquad (8.3)$$

In some cases, a juror may assess that the non-DNA evidence does not favour s, so that on the basis of this evidence no individual is regarded as a more plausible suspect than s. Then $w_i \leq 1$, for all $i \in \mathcal{P}$, and (8.3) leads to

$$P(U \mid E) > \frac{1 - \sum_{i \in \mathcal{P}} R_i}{1 + \sum_{i \in \mathcal{P}} R_i} > 1 - 2 \sum_{i \in \mathcal{P}} R_i. \qquad (8.4)$$

To help motivate the bound (8.4), consider the following simplified example. A person is sampled anonymously and at random from a population $\mathcal{P}$ of size $N + 1$ and is found to have DNA profile Υ. The DNA profiles of the other N individuals are unknown; each is assumed to be Υ independently with probability p, so that the probability that Υ is unique in $\mathcal{P}$ is $P(U) = (1 - p)^N$. A second individual is then drawn at random from $\mathcal{P}$. With probability $1/(N + 1)$, the second individual is the same as the first (call this event G). Now, the second individual is typed and found to have profile Υ (call this observation E). If G holds, E provides no additional information about U, so that

$$P(U \mid G, E) = (1 - p)^N. \qquad (8.5)$$

Otherwise, the two individuals sampled are distinct, yet both have profile Υ, implying that U is false. Substituting (8.5) in (8.1), and noting that the island problem formula (2.1) applies here for $P(G \mid E)$, we obtain

$$P(U \mid E) = P(U \mid G, E)P(G \mid E) = \frac{(1-p)^N}{1 + Np} > 1 - 2Np, \qquad (8.6)$$

which is the bound given by (8.4) in this setting. See Balding (1999) for further details and discussion.

8.1.2 Discussion

When $\theta = 2\%$ (see Section 5.2), simulations indicate that 10 STR loci usually suffice to establish "uniqueness" with probability 99·9%, and 11 loci suffice almost always Balding (1999). Remember, however, that these calculations are based on the crucial assumption that $w_i \leq 1$, which effectively implies that there is no evidence in favour of s. In some cases, there is evidence favouring the defendant. More generally, it is usually not appropriate for the forensic scientist to pre-empt the jurors' assessment of the non-scientific evidence.

Focussing on the directly relevant issue, whether or not the defendant is the source of the crime stain, rather than uniqueness, makes more efficient use of the evidence and, properly presented and explained to the court, can suffice as a basis for satisfactory prosecutions. A calculation of the probability of "uniqueness" may also provide useful information for courts, provided that a satisfactory way is found to explain the underlying assumptions.

8.2 Inclusion/exclusion probabilities

8.2.1 Random man

The concept of a "random man" has at least two meanings:

- informally, it means something like "nobody in particular" or "it could be anyone";

- in scientific usage, it means chosen according to a randomizing device, such as a die, or a computer random-number generator.

Use of the idea of "random man" in the first sense is generally harmless, though it may cause some confusion. Serious errors can arise when "random man" is used in the second sense in assessing weight of evidence. It is important to keep in mind that in any crime investigation "random man" is pure fiction: nobody was actually chosen at random in any population, and so probabilities calculated under an assumption of randomly sampled suspects have no direct bearing on evidential weight in actual cases.

The concept of "random man" is sometimes used to calculate an *inclusion probability* (sometimes reported as its complement, the *exclusion probability*). Jurors in a case in which identification is at issue may be informed that a sequence of tests has been conducted on samples from both crime scene and defendant, and that the defendant was not excluded by these tests from being the source of the crime-scene evidence. Instead of likelihood ratios, the expert witness advises jurors on the probability that a "random man" would not have been excluded by the tests. It

OTHER APPROACHES TO WEIGHT OF EVIDENCE

is important to distinguish two types of inclusion probability, according to whether or not the probability takes into account the evidence profile. In either case, the defendant's profile is not taken into account beyond noting that the defendant has not been excluded by the test(s).

8.2.2 Inclusion probability of a typing system

Consider a system that classifies individuals (here assumed without error) into K distinct classes such that a proportion p_k of individuals is put into class k, and $\sum_{k=1}^{K} p_k = 1$. If classification is independent for distinct individuals, then the inclusion probability for identification using this system is the probability that two random individuals (e.g. innocent suspect and true culprit) are classified together and is given by

$$P(\text{inclusion}) = \sum_{k=1}^{K} p_k^2.$$

This probability is useful for describing the discriminatory power of an identification system: given a choice between implementing one of several classification systems for routine use, all other factors being equal, one would prefer to implement the system with the lowest inclusion probability (equivalently, highest exclusion probability). However, the inclusion probability of the test has little role in assessing the weight of evidence in a crime investigation because it does not take account of any information specific to the particular crime.

8.2.3 Case-specific inclusion probability

Identification: single contributor

In the case of a single contributor to the crime-scene evidence, the probability of inclusion *given* that evidence is the probability that a randomly chosen alternative possible culprit would be assigned to the same class as the true culprit. For DNA profile evidence, each distinct profile corresponds to a class. In the notation introduced above, the inclusion probability is p_k, where k is the class of the crime-scene evidence. In this setting, p_k is numerically equivalent to a likelihood ratio for an unrelated individual (with $\theta = 0$), but the two approaches are conceptually very different.

The rationale underlying the inclusion probability, based on a random alternative suspect, cannot adequately cope with relatives of the defendant among the alternative culprits and also faces difficulties in the presence of missing alleles (partial profiles). The idea of a random alternative suspect can lead jurors to ignore the role of the *number* of possible culprits in evidential assessments, and clear thinking about typing errors and the effect of searches can also be undermined. All of these aspects are readily dealt with in the likelihood ratio framework.

One specific difficulty that "random man" can cause in this setting concerns the argument over which population the man is supposed to have been randomly drawn from (see e.g. Roeder 1994). Since the whole idea of "random man" is a fiction, these arguments can never be resolved. The issue is important, since too broad a definition of the population leads to overstatement of the evidence, because a large population must contain many people sharing little ancestry with s. If we try to avoid this overstatement by specifying the narrowest possible population, we are led to the population consisting of s only, in which the match probability is one.

Identification: multiple contributors

Even more important differences between inclusion probabilities and likelihood ratios arise in the setting of crime-scene DNA profiles with two or more unknown contributors (Section 6.5). Here, the inclusion probability is the probability that a random unknown individual would have both alleles at each locus included among the alleles of the mixed crime-scene profile. Thus, the inclusion probability does not take account of the profile of the defendant, other than noting that it falls within a (usually large) class of profiles. Its advantages include being relatively easy to calculate and explain and that it does not require any assumption about the number of contributors to the mixture.

However, ignoring relevant information can have adverse consequences. Because of the loss of information, the inclusion probability is usually larger than the likelihood ratio, often considerably so. This statistical inefficiency is sometimes seen as a virtue in that use of the inclusion probability is regarded as "conservative" (see Section 6.3). However, conservativeness is not guaranteed (see Exercise 3 below), and by using an appropriate θ-value, we can make the likelihood ratio conservative while still relatively efficient. Another serious disadvantage of the inclusion probability arises in a case involving two co-defendants each accused of being a contributor to the mixed crime-scene profile. The evidence can weigh more heavily against one defendant than the other, whereas the inclusion probability will be the same for both defendants (again, see Exercise 3 below).

Paternity

Another clear distinction between inclusion probabilities and likelihood ratios arises for paternity testing, discussed in Section 7.1. Consider a scenario in which a child's paternal allele at a locus, say allele C, has population proportion p_C. The inclusion probability is the probability that a random man has a copy of the child's paternal allele. Making some simplifying assumptions, this probability is $1 - (1 - p_C)^2 = 2p_C - p_C^2$, irrespective of the genotype of s.

In contrast, the likelihood ratio has already been given at (7.5):

$$R_i = \begin{cases} p_C & \text{if } s \text{ is homozygous for C,} \\ 2p_C & \text{if } s \text{ has just one copy of C.} \end{cases}$$

The inclusion probability lies between these two values. If $p_C > 0.5$, then $R_i > 1$ for s heterozygous, and so this observation *reduces* the probability that s is the father (see Section 7.1.3). This is reasonable because an allele drawn from s is less likely to match the child's paternal allele than is an allele drawn at random in the population. Thus, this evidence points away from s as the father, and towards any unprofiled, unrelated man i.

The inclusion probability approach wrongly counts the evidence in the above scenario as counting against s, just the same as if he had genotype CC. This means that test results that actually favour the defendant (i.e. which decrease his probability of paternity) will wrongly count against him/her in the inclusion probability approach. Although $p_C > 0.5$ is rarely realistic for STR profiling, this extreme scenario highlights the logical problems associated with not answering the relevant question.

As noted in Section 7.1.3, Li and Chakravarti (1988) prefer the inclusion probability as a measure of evidence for paternity, rather than likelihood ratios (= paternity index). We have seen that this method has important weaknesses, which stem from the fact that it ignores pertinent information: it takes no account of the actual profile of s, other than whether or not it is consistent with the alleged family relationships. For further discussion, see Kaye (1989).

Discussion

Inclusion probabilities seem to be attractive as a measure of evidential weight in place of likelihood ratios. There is, however, no theory linking inclusion probabilities with the question of the defendant's guilt. This, in itself, may not be troubling if they satisfied some informal notion of fairness. This is the case in many settings, but we have seen that inclusion probabilities do sometimes lead as astray. One key weakness is that all evidence decreases the inclusion probability and hence counts *against* the defendant. Inclusion probabilities may have some uses in measuring and conveying evidential weight, but they should not be used without checking that they do not conflict with the logical analysis based on likelihood ratios combined using Bayes Theorem.

8.3 Hypothesis testing †

Forensic scientists will be familiar with the scientific convention of assessing the weight of scientific evidence via significance levels and/or p-values. The jury in a criminal case must reason from the evidence presented to it to a decision between the hypotheses:

G : the defendant is the culprit;

I : the defendant is not the culprit.

Within the hypothesis-testing framework, the legal maxim "innocent until proven guilty" would imply that I should be the null hypothesis. The probability of a match under hypothesis I could then be interpreted as a p-value.

But how do we calculate a p-value under hypothesis I, taking into account, for example, the possibility that the true culprit could be a relative of the defendant? The usual answer is to invoke "random man", and assume that hypothesis I implies that s has been chosen randomly in some population of innocent suspects. As we have discussed in Section 8.2, this approach leads to difficulties. For example, since no random sampling really took place, it is impossible to specify the population.

The hypothesis-testing framework faces further difficulties with complications that we have seen are readily handled using the weight-of-evidence formula (3.3):

- How can the p-value, assessing the DNA evidence, be incorporated with the other evidence? What if the defendant produces an apparently watertight alibi? What if more incriminating evidence is found?

- How should the possibility of laboratory or handling error be taken into account?

- What if the defendant was identified only after a number of other possible culprits were investigated and found not to match?

Perhaps the most important weakness is the first: the problem of incorporating the DNA evidence with the other evidence. Hypothesis tests are designed to make accept/reject decisions on the basis of the scientific evidence only, irrespective of the other evidence. Legal decision makers must synthesize all the evidence, much of it unscientific and difficult to quantify, in order to arrive at a verdict.

The rationale behind classical hypothesis testing is based on imagining a long sequence of similar "experiments". Roughly speaking, a p-value addresses the question:

> "If the null hypothesis were true, how often in many similar experiments would I observe data equally or more discordant with the null than the data actually observed?".

For example, in Section 5.4, we discussed hypothesis tests for deviations from Hardy–Weinberg equilibrium (HWE). Here, the null hypothesis of HWE is simple: it specifies precise genotype proportions as functions of allele proportions. The tests, one way or another, answer the question: if we had drawn samples from many populations in HWE, and observed the same allele counts each time, how many times would the genotype counts deviate from the HWE expectations as much as, or more than, the observed sample?

As we have seen, the courtroom setting for DNA identification is very different. The null hypothesis of the defendant's innocence is complex – there are many ways that the defendant can be innocent (i.e. many possibilities for the true culprit) and there can be a different match probability for each alternative. Moreover, imaginary

OTHER APPROACHES TO WEIGHT OF EVIDENCE

repeats of an "experiment" of suspect generation would involve speculation that is inappropriate to the courtroom.

Much confusion has been caused by trying to shoehorn the decision problem faced by jurors into the scientific hypothesis testing framework; see Balding and Donnelly (1995b) for a discussion. The second report on DNA evidence of the US National Research Council (1996) adopted this framework for most of its statistical deliberations, and consequently, some of its conclusions are flawed, and much of its discussion is misspent tackling the wrong issues; see Section 9.4.

8.4 Other exercises

Solutions start on page 172.

1. In the original island problem of Section 2.2.1, we calculated a probability of 50% for the defendant's guilt. Assuming no evidence other than the Υ evidence, what is the probability that s is the only Υ-bearer on the island?

2. (a) What is the inclusion probability for the identification test used in the island problem? That is, what is the probability that two individuals will be concordant for a binary trait when the population proportions having the two states of the trait are p and $1 - p$? What value of p is optimal (minimizes the inclusion probability)?

   (b) Assume the following population proportions for the four main blood groups:

Blood group:	O	A	B	AB
Population proportion:	46%	42%	9%	3%

   Assume that the blood groups of distinct individuals are independent. When considering the identification of the source of a blood sample drawn from a single individual, what are

   i. the inclusion probability of the typing system based on these blood groups?
   ii. the case-specific inclusion probabilities for each of the four possible blood groups?

   Briefly explain the uses of the probabilities calculated in (i) and (ii).

3. Consider a single-locus genetic typing system at which a mixed DNA sample recovered from a crime scene has profile ABD, while a man s suspected of being a contributor to the sample has genotype BD. The population allele proportions are $p_A = 0.2$, $p_B = 0.1$, and $p_D = 0.15$, and genotype proportions equal Hardy–Weinberg values (Section 5.4). We assume here that none of the actual or possible contributors to the crime stain is related and that $\theta = 0$.

(a) What is the value of the likelihood ratio comparing the hypothesis that s and an unknown man were the (only) contributors to the mixture, with the hypothesis that two unknown men were the sources of the crime-scene sample?

(b) What is the inclusion probability?

(c) Repeat (a) and (b) with $s \equiv BB$.

(d) Outline (without any calculations) how your answers in (a) and (b) would be altered if it were known that the mixture had three contributors?

4. †Suppose that s has been charged with an offence because of being the only one of 1000 individuals included in a database whose DNA profile matched the crime-scene profile. Consider the hypothesis:

H_0: The profiles recorded in the database were drawn from individuals chosen at random in the total population of 10 000 mutually unrelated possible culprits.

Assume that each innocent individual has probability 10^{-6} of having a matching profile, and that the profiles of different individuals are independent. The DNA profiles of the 9000 individuals not in the database are unknown.

(a) What is the significance level of the test that rejects H_0 if at least one match is observed of the crime-scene profile with a database profile?

(b) What is the relevance of your answer to (a) for the strength of the case against s?

9

Issues for the courtroom

My hope is that forensic scientists reading this book will, if they have persevered this far, have gained useful insights about how to quantify the weight of DNA evidence, and that these insights will help with many aspects of preparing statements and presenting them in court. If any such reader was hoping that I would be able to prescribe a formulaic approach to reporting DNA evidence in court, that satisfies the needs of jurors and the demands of judges in every case, then s/he will be disappointed. I have no magic formula to overcome the difficult issues that arise in presenting complex scientific evidence to non-expert judges and juries. However, I believe that a sufficiently deep understanding of the principles can help an expert witness to make well-informed judgements about what a clear-headed juror needs to know in order to perform his or her task. It must be up to the individual forensic scientist, reflecting their context, to find good solutions to the problem of satisfying the (partly contradictory) goals of, for example, clarity, precision, fairness, exhaustiveness, and simplicity.

9.1 Bayesian reasoning in court

The principles that I have set out here are based on the use of Bayes Theorem to assess evidence. Arguments over the acceptability of explicit Bayesian reasoning in court have been the subject of academic debate at least since the critique of Tribe (1971) provoked by Finkelstein and Fairley (1970) (see also the authors' response, Finkelstein and Fairley 1971). The prospect of formalized reasoning with numbers seems not to have been welcomed by the courts. Indeed, we will see in Section 9.3.3 that a UK Court of Appeal seems to have ruled Bayes Theorem to be inadmissible as part of evidence. While it is regrettable that rational reasoning is officially prohibited from UK courtrooms, it is also clear that most of the statistical and population-genetics analyses described in this book are not appropriate for direct presentation to juries.

The critics of Bayesian reasoning in court have some support. There is a substantial academic literature pointing to the conclusion that lay jurors are prone to confusion when presented with evidence in the form of numbers. In particular, forms of words attempting to encapsulate likelihood ratios have been highlighted for their potential to sow confusion in jurors' minds; see, for example, Thompson (1989) and Koehler (1993b, 1996, 2001).

These criticisms should be treated as encouragements to be cautious. On the other hand, now that DNA evidence is routinely quantified numerically, and there may often be substantial additional evidence of a non-numeric form, there seems to be no better proposal on offer to deal with the complexities that this raises than to start with a logical analysis via the weight-of-evidence formula (3.3), or similar version of Bayes Theorem. Many scientists – we will see below (see Section 9.4) even members of a US National Research Council panel – have become muddled about what scientific issues are relevant to a juror's decision, and logical reasoning has a central role to play in clarifying these issues, even if such an analysis is not directly presented in court. My own attempts to convey evidential weight in court have been based on this formula, but I rarely utter the words "likelihood ratio" or "Bayes Theorem" in court. In the interests of jurors unfamiliar with formalized reasoning, I believe that the intuition given by the mathematical formalism can, and should, be conveyed in ways that avoid its explicit use. I see the theory developed in this book as crucial for good understanding by the forensic scientist, and not as a prescription for how to convey this understanding in court.

9.2 Some fallacies

9.2.1 The prosecutor's fallacy

The "Prosecutor's Fallacy" (Thompson and Schumann 1987) is a logical error that can arise when reasoning about DNA profile evidence. You will be aware of the error in elementary logic of confusing "A implies B" with "B implies A". For example, if A denotes "is a cow" and B denotes "has 4 legs", then (ignoring rare anomalies) A implies B, but B does not imply A. The prosecutor's fallacy is similar to this logical error, but is in terms of probabilities.

The fallacy consists of confusing $P(A \mid B)$ with $P(B \mid A)$.[1] If it is accepted that there is a 90% chance that the culprit is very tall, it does not follow that every very tall man has 90% probability of being the culprit. The correct way of obtaining $P(A \mid B)$ from $P(B \mid A)$ is given by the appropriate version of Bayes Theorem, for example, the weight-of-evidence formula (3.3).

Transcripts of court cases have often recorded statements that indicate that the match probability is being confused with the probability that the defendant is

---

[1] The prosecutor's fallacy is a particular case of the "error of the transposed conditional".

ISSUES FOR THE COURTROOM

innocent. For example:

- "I can estimate the chances of this semen having come from a man other than the provider of the blood sample ... less than 1 in 27 million";
- "The FBI concluded that there was a 1 in 2600 probability that the semen ... came from somebody other than Martinez".

In these quotations, an expert witness has made a statement about the probability that the defendant is not the source of the crime profile. Such statements lie outside the domain of an expert witness, are logically distinct from a match probability or likelihood ratio, and appear to be instances of the prosecutor's fallacy. A correct version of these statements should involve a conditional statement of the form "*if* the defendant were not the source, then the probability ...". The fallacy can be extremely detrimental to defendants, and has led to successful appeals in the UK (Sections 9.3.1 and 9.3.2).

In addition to clear instances of the fallacy, there are many ambiguous phrases that could suggest the fallacy to jurors. For example, a sentence of the form

"The probability that an innocent man would have this profile is 1 in 1 million"

is ambiguous because it is unclear whether this refers to a match probability for a particular man (probably OK), or to the probability that there exists (anywhere) an innocent, matching man (probably fallacious).

For further examples and discussion of the fallacy, see Koehler (1993b), Balding and Donnelly (1994), Evett (1995), and Robertson and Vignaux (1995). Good (1995) exposes an instance of a similar error of reasoning with probabilities. In the US trial of O.J. Simpson, a member of the defence team pointed out that few wife-batterers go on to murder their wives, drawing the conclusion that Simpson was *a priori* unlikely to be guilty. Good points out that this ignores the fact that Simpson's ex-wife had been murdered, and the available data suggest that a high proportion of murdered women who have been battered by their husbands were in fact murdered by their husbands.

9.2.2 The defendant's fallacy

Another error of logic that can arise in connection with DNA evidence usually favours the defendant and is consequently dubbed the "Defendant's Fallacy". Suppose that a crime occurs in a nation of 100 million people and a profile frequency is reported as 1 in 1 million. The fallacy consists of arguing that, since the expected number of people in the nation with a matching profile is 100, the probability that the defendant is the culprit is only 1 in 100, or 1%.

This conclusion would be valid only if, ignoring the DNA evidence, every person in the nation is equally likely to be the culprit. (In the notation of the weight-of-evidence formula, each w_i is equal to one.) In practice, such an assumption is rarely reasonable. Even if there is little or no directly incriminating evidence beyond the DNA profile match, there is always background information presented in evidence, such as the location and nature of the alleged offence, which will make some individuals more plausible suspects than others.

A closely related fallacy consists of arguing that, since it is expected that many people in the nation share the profile, the DNA evidence is almost worthless and should be ignored. This is clearly false: even if the defendant's fallacy argument is accepted, the DNA evidence is very powerful in reducing the number of possible culprits from 100 million to just 100.

9.2.3 The uniqueness fallacy

Consider a country of population n, and a DNA profile with match probability for an unrelated, alternative culprit calculated to be less than $1/n$. Ignoring relatives, coancestry, and any other evidence, this implies a mathematical expectation of less than one for the number of matching individuals in the population. The fallacy is to conclude that an expectation less than one means that no such individual exists. In the UK case R. v. Gary Adams (Section 9.3.4), a match probability of 1 in 27 million had been reported and the judge concluded:

> "... I should think that there are not more than 27 million males in the United Kingdom, which means that it is unique."

When the expectation of a count is very small, it is approximately equal to the probability that the count is not zero. Thus, the match probability would have had to have been several orders of magnitude smaller than 1 in 27 million in order for the judge's conclusion to be reasonable.

9.3 Some UK appeal cases

9.3.1 Deen (1993)

$Deen^2$ was the first conviction based primarily on DNA evidence to come before a UK appeal court. The judgement upholding the appeal is notable for establishing the prosecutor's fallacy as an important issue in the presentation of DNA evidence.

At the trial, an expert witness agreed with the statement:

> "the likelihood of [the source of the semen] being any other man but Andrew Deen is 1 in 3 million",

a clear example of the fallacy. There were further ambiguous statements, which could have been interpreted as instances of the fallacy.

The appeal judgement explained the fallacy at length and stated explicitly the judges' view that it was not necessary for them to assess how much difference the

---

[2]*The Times*, January 10, 1994.

misrepresentation of the DNA evidence would have made. In addition to the prosecutor's fallacy, there was a separate ground for appeal concerning the presentation of evidence in connection with missing bands.

9.3.2 Dalby (1995)

The prosecutor's fallacy was again one ground for a successful appeal in *Dalby*, even though there was not an explicit instance of the fallacy documented at the trial. The prosecution evidence included ambiguous statements in connection with the DNA evidence, such as:

> "The chance of occurrence of the profile obtained from the semen stain on the pants is estimated to be less than 1 in 2.7 million."

This failure to present the DNA evidence clearly, and to warn the jury against making the prosecutor's fallacy, led the appeal court to take the view that the danger that the jury fell into the fallacy was unacceptably high.

In addition, on the basis of a reported profile frequency of 1 in 2.7 million, the trial judge in his summing up said:

> "... there are 837 000 males in the age group 30 to 44 years which would mean that statistically there would be no other male [in the region] other than the defendant with the same DNA profile – though of course the one white male [expected] in England and Wales might live in that area."

This statement was criticized as misleading by the higher court, and formed another of the grounds for upholding the appeal.

9.3.3 Adams (1996)

In April 1996, the Court of Appeal in London overturned the conviction for rape of Denis Adams.[3] The case was unusual in that, although the prosecution case was supported by a DNA profile match linking the defendant with the crime, for which a match probability of 1 in 200 million was reported, the defence case also had strong support, from an alibi witness who seems not to have been discredited at trial, but most importantly from the victim, who stated in court that the defendant did not look like the man who attacked her nor did he fit the description that she gave at the time of the offence.

Thus, jurors were faced with the task of weighing the strongly incriminating DNA evidence against substantial non-scientific evidence pointing in the other direction. Experts for prosecution and defence agreed that the logically satisfactory way to do this was via Bayes Theorem (in effect, the weight-of-evidence formula (3.3)). The defence presented numerical illustrations of the theorem applied

---

[3][1996] 2 Cr.App.R. 467. See also *The Times*, 9 May, 1996; *New Scientist*, 8 June, 1996, and 13 December, 1997.

sequentially to each of the principal pieces of evidence. The jury was reminded repeatedly that they should assess their own values. In his summing up, the trial judge reminded jurors that they were not obliged to use Bayes Theorem.

An appeal was launched for reasons other than the principle of using Bayes Theorem, but although not raised by the appellant, in upholding the appeal the court expressed strong reservations as to whether such evidence should have been allowed to go before the jury. They added that, as they had not heard argument on the matter, their views should be interpreted as provisional.

It is difficult for my non-legal mind to find much of substance in the appeal court judgement. The judges express concern that in advising jurors of the accepted, rational method of reasoning with probabilities, the Bayes Theorem evidence

> "... trespasses on an area peculiarly and exclusively within the province of the jury, namely the way in which they evaluate the relationship between one piece of evidence and another."

The substantive paragraph seems to the one which says:

> "... the [Bayes Theorem] methodology requires ... that items of evidence be assessed separately according to their bearing on the accused's guilt, before being combined in the overall formula. That in our view is far too rigid an approach to evidence of the nature which a jury characteristically has to assess, where the cogency of ... identification evidence may have to be assessed ... in the light of the strength of the chain of evidence of which it forms part."

This is somewhat unclear, but it seems that the judges have misunderstood the sequential application of Bayes Theorem to different items of evidence (Section 3.3.1). They go on to say:

> "... the attempt to determine guilt or innocence on the basis of a mathematical formula, applied to each separate piece of evidence, is simply inappropriate to the jury's task. Jurors evaluate evidence and reach a conclusion not by means of a formula, mathematical or otherwise, but by the joint application of their individual common sense and knowledge of the world to the evidence before them."

The issue of how common sense and knowledge of the world equip jurors to understand the weight of DNA evidence, presented as a match probability, was not elaborated by the judges.

The successful appeal led to a retrial at which Bayes Theorem was introduced again (the higher court's opinion was not binding and the retrial judge chose to ignore it). In fact, the court went further and, with the collaboration of prosecution and defence, distributed to the jury a questionnaire that guided them through the application of Bayes Theorem, together with a calculator. Adams was convicted

ISSUES FOR THE COURTROOM 151

again, and appealed again. The Court of Appeal, in dismissing the second appeal,[4] took the opportunity to reinforce its earlier opinion:

> "... expert evidence should not be admitted to induce juries to attach mathematical values to probabilities arising from non-scientific evidence ..."

Clearly, jurors may be confused by unfamiliar mathematical formalism, and indeed, it seems that the judge did make errors in his summing up of the Bayes Theorem evidence. Nevertheless, once DNA evidence is presented in terms of probabilities, it seems hard to sustain the argument that jurors should be *prevented* from hearing an explanation of the accepted method of reasoning with probabilities to incorporate it with the non-DNA evidence. The appeal court judgements give no hint as to how jurors might otherwise be guided in this task.

9.3.4 Doheny/Adams (1996)

In July 1996, the Court of Appeal in London considered two related appeals[5] against convictions based to varying extents on DNA evidence. The court upheld one appeal and dismissed the other; the prosecutor's fallacy was an issue in both cases, but there were additional issues concerning the calculated match probabilities.

The appeals are important because the Court of Appeal took the opportunity that they offered to give directions as to how DNA evidence should be presented at trial:

> "[the scientist] will properly explain to the Jury the nature of the match .... He will properly, on the basis of empirical statistical data, give the Jury the random occurrence ratio – the frequency with which the matching DNA characteristics are likely to be found in the population at large. Provided that he has the necessary data, and the statistical expertise, it may be appropriate for him then to say how many people with the matching characteristics are likely to be found in the United Kingdom – or perhaps in a more limited relevant sub-group, such as, for instance, the Caucasian sexually active males in the Manchester area.
>
> This will often be the limit of the evidence which he can properly and usefully give. It will then be for the Jury to decide, having regard to all the relevant evidence, whether they are sure that it was the Defendant who left the crime stain, or whether it is possible that it was left by someone else with the same matching DNA characteristics.

---

[4] [1998] 1 Cr.App.R 377
[5] [1997] 1 Cr App R 369. See also *The Times*, 14 August, 1996. Note that this "Adams" is different from the one discussed above.

The term "random occurrence ratio" introduced by the court appears to be a synonym for match probability. This novel coinage is an unwelcome addition to the many terms already available: its unfamiliarity could confuse.

The court then suggested a model for summing up DNA evidence, which implies assessing the DNA evidence before taking any other evidence into account:

> "Members of the jury, if you accept the scientific evidence called by the Crown, this indicates that there are probably only four or five white males in the United Kingdom from whom that semen stain could have come. The Defendant is one of them. The decision you have to reach, on all the evidence, is whether you are sure that it was the Defendant who left that stain or whether it is possible that it was one of the other small group of men who share the same DNA characteristics."

This approach is not illogical, but it encounters several difficulties. The court does not seem to have considered the effect of relatives on evidence, nor the problem of integrating the DNA evidence with the other evidence. However, an expert can take relatives into account when calculating the expected number of matches and can emphasize that the calculation does not take the non-DNA evidence into account.

With current match probabilities, the expected number of matching individuals in the UK is fractional, usually a small fraction. This may be difficult for jurors to interpret. In particular, it is important to clarify that probabilities refer to unknowns: they summarize the information about something not completely known. Therefore, a statement such as:

> "The expected number of individuals in the UK sharing the crime-scene profile is 0·01"

may appear to a juror to conflict with the fact that the defendant has the profile. The phrase "other than the defendant", or similar, may help avoid this possible confusion. Alternatively, when expectations are small, it may be preferable to use probabilities; for example:

> "There is only one chance in a hundred that any other person in the UK shares the crime-scene profile."

Although not without merit, the court's prescription for the presentation of DNA evidence has had unsatisfactory consequences in its attempts to limit what a forensic scientist may comment on. In addition to the phrase "That will often be the limit ..." quoted above, the court went on to say

> "The scientist should not be asked his opinion on the likelihood that it was the defendant who left the crime stain, nor when giving evidence should he use terminology which may lead the jury to believe that he is expressing such an opinion".

ISSUES FOR THE COURTROOM 153

It seems clear from the context that the Court of Appeal was trying to restrain expert witnesses from any danger of falling into the prosecutor's fallacy, but in some UK courts, these words seem to have been used to straightjacket forensic scientists. This is unfortunate: particular DNA cases have their own complexities, and DNA profiling technology has advanced rapidly since the judgement. For a further discussion of the its implications, see Lambert and Evett (1998).

9.3.5 Watters (2000)

The appellant had been convicted of burglary on the basis of STR profiles obtained from cigarette ends obtained at the crime scenes, together with reports of similarities between these crimes.

A forensic expert reported matches at six STR loci. The match probability for an unrelated individual was said to be 1 in 86 million, whereas that for a brother was 1 in 267. It was reported in evidence that the defendant had two brothers and that one of these had been arrested in connection with the offences, but released without charge and without a DNA sample having been taken from him (or the other unarrested brother). The forensic scientist agreed that her results did not prove that the defendant was the source of the crime-scene DNA and also that DNA evidence should not be used in isolation and without other supporting evidence. The supporting evidence provided by the prosecution included that the defendant was a smoker and lived in the general locality of the burglaries.

The defence requested the trial judge to, in effect, rule that there was no case to answer because the brothers had not been excluded. He rejected this argument and invited the jury to consider the 1 in 267 match probability for brothers. His arguments for rejecting the defence request include the existence of the supporting evidence about locality, and because the defence had not supplied the names, addresses, and dates of birth of the brothers. The latter argument would seem to contravene the principle that a defendant does not have to prove anything.

The appeal court found that ruling to be wrong, on the basis of a mixture of valid and misguided reasoning. The court drew attention to the forensic scientist's admission that a DNA profile match does not constitute proof of identity and contrasted DNA evidence with fingerprint evidence where certain identification is routinely reported in court. This contrast is spurious, and it is regrettable that the attempt to assess probabilities for error in the case of DNA profile evidence was interpreted by a senior court as implying that DNA evidence is inferior to fingerprint evidence: comparable difficulties for fingerprint evidence have routinely been swept under the carpet (see Section 4.6).

The court was on stronger ground in asserting that, although the DNA evidence was very strong in excluding a source of the DNA unrelated to the defendant, the existence of unexcluded brothers, one of whom had been suspected of the offences, cast doubt on whether the correct brother was in court. In my view, the court should be congratulated for taking the issue of the unexcluded brothers seriously, which courts in many earlier trials failed to do. However, its interpretation of "certainty" is doubtful: two brothers each with match probability less then 1 in 200 could, in the absence of other evidence and other possible culprits, correspond to a posterior probability of guilt of over 99%, which a reasonable juror may accept as "certainty" within the meaning of the law.

Whatever the merits of that argument, it seems absurd to argue, as the Court of Appeal did in its judgement, that further STR testing, which resulted in the defendant continuing to match but now with a match probability for a brother of 1 in 29 000, was still insufficient for a conviction because this much smaller match probability still does not suffice to eliminate the possible brother. On this basis they did not order a retrial. The implication of the court's argument seems to be that any uncertainty quantified numerically fails to achieve "proof". Uncertainty prevails for all forms of evidence, and the attempt to quantify should be seen as a sign of the strength of DNA evidence, not as a weakness.

9.4 US National Research Council reports

The report on DNA evidence of the US National Research Council (1992) was intended to settle some of the controversy surrounding DNA evidence. The report devoted almost no space to weight-of-evidence issues and espoused no principles for assessing evidence. In response to the perceived problem of population heterogeneity, it proposed an *ad hoc* solution, the "ceiling principle", which involves using the largest frequency for each DNA profile band, assessed from a diverse range of populations. The report suffered a hostile reception from many sides. In particular, the ceiling principle was criticized, many taking the view that it was too generous to defendants and others complaining that it was a poorly considered response to a poorly specified problem.

Following pressure from, principally, the FBI, a new committee was formed and a new report issued (National Research Council 1996). This time there was a (very brief) discussion of evidential weight. Unfortunately, the report adopted the principle that weight of evidence is measured in terms of an arbitrary and unjustified assumption that the defendant was sampled randomly in some population (see the discussion of "random man" in Section 8.2.1). Consequently, the second report is also flawed in many important respects. Perhaps, the most astonishing feature is the serious error on the issue of a defendant identified by a database search, mentioned already in Section 3.4.5. The report recommends that the weight of DNA evidence in this setting be measured by the profile frequency multiplied by the size of the database, indicating much weaker evidence than in a no-search case.

The rationale for this recommendation is clearly flawed. The report compares the DNA search scenario to tossing 20 coins repeatedly so that eventually it becomes likely that the outcome "all heads" will have occurred. This analogy is misleading: in the DNA search case, we know that somebody, somewhere, has the profile. Moreover, as we noted in Section 2.3.4, the relevant question is not "what is the probability of observing a match?" but "given that I have observed a match, how strong is the evidence against this individual?". The correct analysis

ISSUES FOR THE COURTROOM   155

of the database search scenario was discussed in Section 3.4.5; the number of profiles in the database is essentially irrelevant to evidential weight, and the NRC's recommendation is similar to suggesting that a good answer be multiplied by an arbitrary number.

Aside from the database search blunder, which gives a large and unwarranted benefit to some defendants, the report otherwise errs consistently in favour of prosecutions. Although the committee that prepared the report was chaired by one eminent population geneticist and included several others, failure to adopt appropriate weight-of-evidence principles means that the population-genetics theory was misapplied. Because it considered only profile proportions for "random man", the report fails to address the key population genetics issue of the correlations between the DNA profile of the defendant and those of other possible culprits. Instead, the report gives undue attention to the less important issue of within-person genetic correlations, and recommendations substantially favouring prosecutions are the result.

Recommendation 4.3, concerning small and isolated groups such as Native American tribes, for which little relevant data is available, is particularly troubling and could lead to miscarriages of justice. Similarly, troubling is the report's recommendation that relatives be considered only when *there is evidence* that they are possible culprits. We saw in Section 3.4.3 how this approach can be misleading. Moreover, this assumption in effect reverses the burden of proof: it should be for the prosecution to prove that the defendant committed the offence, and hence that his relatives did not, rather than for the defence to show that relatives are plausible suspects. Similarly unfair to defendants is the recommendation that subpopulation issues be taken into account only if *all* the alternative possible culprits are from the same subpopulation as the defendant. See Balding (1997) for further discussion and criticism of the report.

9.5 Prosecutor's fallacy exercises

Smith has a genetic type that matches that of blood found at a crime scene. The likelihood ratio for an individual unrelated to Smith is reported to be 1 in 1000. The statements below are adapted from Evett (1995). Discuss whether each is probably OK, probably an instance of the Prosecutor's Fallacy, or ambiguous.

1. The probability that the profile would be of this type if it had come from someone other than Smith is 1 in 1000.

2. The chance that a man other than Smith left blood of this type is 1 in 1000.

3. The probability that someone other than Smith would have blood of this type is 1 in 1000.

4. The evidence is 1000 times more likely if Smith were the source of the blood than if an unrelated man were the source.

5. It is 1000 times more probable that Smith is the source of the blood than that an unrelated man is the source.

6. The chance of a man other than Smith having the same blood type is 1 in 1000.

10

Solutions to exercises

Section 2.4, page 20

1. (a) Manchester:

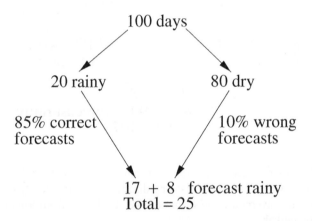

Out of every 25 rainy days, on average 17 had been correctly forecast as rainy, whereas 8 had been wrongly forecast to be dry. The probability of rain given a forecast of rain is equal to the proportion of rainy days on which rain had been forecast, which is $17/25 = 68\%$.

In Chapter 3, we will use mathematical formalism and Bayes Theorem (3.9) to compute this answer more concisely. Here is a preview of a more formal derivation. Let

$$W \equiv \text{"rainy"}$$
$$\overline{W} \equiv \text{"dry"}$$
$$F \equiv \text{"rain forecast"}.$$

Weight-of-evidence for Forensic DNA Profiles   David Balding
© 2005 John Wiley & Sons, Ltd   ISBN: 0-470-86764-7

We seek $P(W \mid F)$. Now

$$P(W) = 0.2, \qquad P(F \mid W) = 0.85,$$
$$P(\overline{W}) = 0.8, \qquad P(F \mid \overline{W}) = 0.1,$$

and by Bayes Theorem we have

$$P(W \mid F) = \frac{P(F \mid W)P(W)}{P(F \mid W)P(W) + P(F \mid \overline{W})P(\overline{W})}$$
$$= \frac{0.85 \times 0.2}{0.85 \times 0.2 + 0.1 \times 0.8} = \frac{0.17}{0.25} = 0.68.$$

(b) Alice Springs:

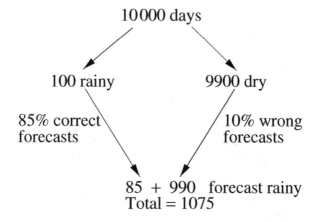

So, after an accurate forecast of rain, there is only 85/1075, just under 8%, chance of rain. The strong diagnostic information (forecast of rain) cannot overcome the even stronger background information that rain is unlikely.

Using Bayes Theorem:

$$P(W \mid F) = \frac{0.85 \times 0.01}{0.85 \times 0.01 + 0.1 \times 0.99} = \frac{0.0085}{0.1075} \approx 0.079.$$

2. (a) Because of the extremely strong background information that the condition is very rare, after an accurate diagnosis of the condition there is still only 992/20 990, just under 5%, chance of having it. We derive this first using Bayes Theorem:

$$P(W \mid F) = \frac{0.992 \times 0.0001}{0.992 \times 0.0001 + 0.002 \times 0.9999}$$
$$= \frac{0.0000992}{0.0000992 + 0.0019998} = \frac{992}{20\,990} \approx 0.047,$$

and now represent the derivation using a diagram:

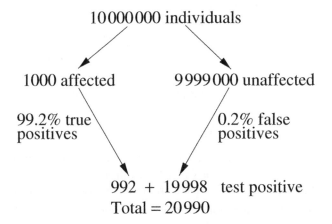

(b) Using Bayes Theorem:

$$P(W \mid F) = \frac{0{\cdot}992 \times 0{\cdot}004}{0{\cdot}992 \times 0{\cdot}004 + 0{\cdot}002 \times 0{\cdot}996}$$
$$= \frac{0{\cdot}003968}{0{\cdot}003968 + 0{\cdot}001992} = \frac{3968}{5960} \approx 0{\cdot}67.$$

Thus, although the background evidence implies that the person tested is unlikely to have the condition (probability $1/250 = 0{\cdot}004$), the diagnostic information from the accurate test result is so strong that it overcomes this and results in a 2/3 probability that the person does have the condition.

3. (a) Use of the island problem formula (2.1) gives

$$P(G \mid E) = \frac{1}{1 + N \times p} = \frac{1}{1 + 1000 \times 5 \times 10^{-6}} = \frac{1}{1{\cdot}005} \approx 0{\cdot}995.$$

Thus, although there are 1000 alternative possible culprits on the island, because they are all unrelated to s, and the match probability in this case is very small, under the assumptions of the island problem the observation that s has Υ reduces the total probability of guilt for all alternatives from 1000/1001 to 1/201. Conversely, the probability that s is the culprit has increased from 1/1001 to 200/201 or about 99·5%.

(b) Since the islanders are initially equally under suspicion, but have different levels of relatedness to s and hence different match probabilities, formula (2.7) applies to calculate the probability of guilt:

$$P(G \mid E) = \frac{1}{1 + \sum_{i=1}^{N} r_i}$$

$$= \frac{1}{1 + 0.015 + 20 \times 2 \times 10^{-4} + 979 \times 5 \times 10^{-6}}$$

$$= \frac{1}{1 + 0.015 + 0.004 + 0.004895}$$

$$= \frac{1}{1.023895} \approx 0.977.$$

Under these assumptions, despite the Υ match there remains probability of almost 2.5% that s is not the culprit – about 5 times higher than when we assumed that there are no relatives of s on the island.

(c) Under the assumptions of the island problem, the observation that b does not have Υ excludes him from suspicion, and hence the corresponding term vanishes from the match probability formula, leaving

$$P(G \mid E) = \frac{1}{1 + 0.004 + 0.004895} = \frac{1}{1.08895} \approx 0.991.$$

(d) The juror should use whatever background information is available to assess plausible ranges for the numbers of relatives in each category and the probability of different numbers in each range. This might entail using everyday knowledge to assess typical numbers of siblings, nephews/uncles, cousins, and so on. In principle, the weight-of-evidence formula should then be used with each plausible set of numbers of relatives, and the resulting values averaged according to the probabilities of each set. In practice, at the cost of an error in favour of s, it is often satisfactory to consider only an upper bound on the plausible number of relatives in each category, for example, 10 siblings and 100 cousins.

Section 3.6, page 40

1. (a) The juror needs to assess the probability that s would have glass fragments with refractive index matching the crime-scene fragments if in fact s is innocent and i is the true culprit, and also the same probability if s did commit the crime.

   (b) Different individuals i might have different probabilities of carrying glass fragments of a specific type on their clothing according to their trade, hobbies, and lifestyle. For individuals i that live in the same dwelling as s, or other close associates of s, the relevant probabilities may be affected by the fact that s has such fragments and the degree

SOLUTIONS TO EXERCISES   161

of contact between i and s. This is because of the possibility of cross contamination by body contact or exchange of clothing.

(c) Some possible answers:

i. Surveys of the frequency with which glass fragments are found on the clothing of individuals in the general population;

ii. Surveys of the relative frequencies of different refractive indices on clothing;

iii. Evidence about the frequency with which glass fragments are detected on the clothing of individuals who have recently broken a window, after a time delay corresponding to that between the crime and the apprehension of the defendant and the seizure of his/her clothing;

iv. Estimates of the error distribution for measurements of glass fragments obtained from clothing.

2. (a) Including brothers:

$$P(G \mid E) = \frac{1}{1 + 2/267 + 10^5/87 \text{ million}}$$

$$= \frac{1}{1 + 2/267 + 1/870} \approx 0.991$$

(b) Excluding brothers:

$$P(G \mid E) = \frac{1}{1 + 1/870} \approx 0.999.$$

Thus, under our assumptions, the overall case against s is very strong even if the brothers are included and is overwhelming if they are excluded.

3. Let us write

E_d for the DNA evidence

E_o for the background information

(i.e. location of crime, local adult male population)

E_a for the alibi evidence

E_v for the victim's evidence

(a) If we assume that $w_i = 1$ for all 1.5×10^5 men in $\mathcal{P}$, taken here to be the population of men between 15 and 60 within 15 km of the crime scene, and $w_i = 0$ for all other individuals, and that the likelihood ratios for E_d are $R_i = 5 \times 10^{-9}$ (= 1 over 200 million) for all i, then using the weight-of-evidence formula (3.3), the probability of the defendant's guilt would be

$$P(G \mid E_d, E_o) = \frac{1}{1 + \sum_{i \in \mathcal{P}} R_i} = \frac{1}{1 + N \times R_i}$$
$$= \frac{1}{1 + 1.5 \times 10^5 \times 5 \times 10^{-9}}$$
$$= \frac{1}{1 + 7.5 \times 10^{-4}} = \frac{4000}{4003}$$
$$\approx 0.999.$$

(b) Using equations (3.5) and (3.6), the joint likelihood ratio for two items of evidence is the product of the two likelihood ratios, but the second ratio should take into account the first piece of evidence. Here, it seems natural to assume that the alibi evidence and DNA evidence are independent, and so the overall likelihood ratio is the product of the likelihood ratio for the alibi evidence (4) and that for the DNA evidence (5×10^{-9}). Thus, we have

$$P(G \mid E_a, E_d, E_o) = \frac{1}{1 + 1.5 \times 10^5 \times 4 \times 5 \times 10^{-9}}$$
$$= \frac{1}{1 + 3 \times 10^{-3}} = \frac{1000}{1003}$$
$$\approx 0.997.$$

(c) Reasoning the same way as in (b) and assuming that the victim's evidence is independent of both alibi and DNA evidence, we have

$$P(G \mid E_v, E_a, E_d, E_o) = \frac{1}{1 + 1.5 \times 10^5 \times 10 \times 4 \times 5 \times 10^{-9}}$$
$$= \frac{1}{1 + 3/100} = \frac{100}{103}$$
$$\approx 0.97.$$

(d)

$$P(G \mid E_v, E_a, E_d, E_o) = \frac{1}{1 + 1.5 \times 10^5 \times 10 \times 4 \times 5 \times 10^{-8}}$$
$$= \frac{1}{1 + 3/10} = \frac{10}{13}$$
$$\approx 0.77.$$

(e) There are many unreasonable assumptions, among them that s has no relatives in the population $\mathcal{P}$ and that nobody outside $\mathcal{P}$ is a possible culprit. None of the probability assignments can be regarded as precise, but in particular, there seems little basis for assigning $R_i = 4$ for the alibi evidence, rather than, say, 3 or 5 because there is little available data on the reliability of alibi witnesses, and an appropriate value will be specific to the particulars of the case. (Nevertheless, jurors can consider ranges of plausible values, and making such subjective judgements on the veracity of witnesses is what jurors are there to do.) The calculations in (a) through (d) should thus be regarded as providing an aid to clear thinking about the problem rather than as providing a definitive answer to it.

(f) The question is open-ended and gives scope for individual answers. In both trials of Adams (Section 9.3.3), the defence guided jurors through a calculation similar to those made above, allowing jurors to fill in probability assignments. In both appeals, this use of mathematical formalism was severely criticized by the judges, but they could offer no reasonable alternative for conveying the relevant insights about combining different items of evidence.

4. (a) Formula (2.5) applies (see the discussion in Section 3.4.5). The database search eliminates from suspicion the $k = 20$ individuals found not to have Υ, and so

$$P(G \mid E) = \frac{1}{1 + (N - k)p} = \frac{1}{1 + 80 \times 0 \cdot 01} = \frac{5}{9} \approx 0 \cdot 56.$$

The effect of the search is thus to increase the probability that s is the culprit from 50% to about 56%.

(b) Your *a priori* assessment can be formalized as: $w_i = 1$ for every i whose Υ-state is recorded in the database, and $w_i = 0 \cdot 1$ for all other islanders. Of the former group, all but s are eliminated from suspicion, and so, other than s only the 80 individuals with $w_i = 0 \cdot 1$ contribute to the denominator of the weight-of-evidence formula (3.3):

$$P(G \mid E) = \frac{1}{1 + \sum_{i=1}^{80} w_i R_i} = \frac{1}{1 + 80 \times 0 \cdot 1 \times 0 \cdot 01} = \frac{1}{1 \cdot 08} \approx 0 \cdot 93.$$

5. (a) When most of the individuals in the database are not plausible alternative suspects for a particular offence, their elimination due to non-matching with the crime-scene profile has no effect on the case against s. However, because some members of the database may have been considered as (perhaps remotely) possible culprits, and because observation of the non-matches confirms the belief that the profile is rare, the database search always has at least a small effect of strengthening the case against the one individual who does match.

(b) The effect on the probability analysis of information such as that s was ill and far from the crime scene is manifested through large values of w_i for at least some i. Remember that w_i is defined (page 24) as the weight of the other (i.e. non-DNA) evidence against i relative to its weight against s. Since the non-DNA evidence here makes it very unlikely *a priori* that s committed the crime, there are typically many men for whom there is no such exculpatory evidence and so for whom a very large value of w_i is appropriate. For illustration, suppose that the possible suspects include s, thought very unlikely to have committed the crime, and say 10 000 men with $w_i = 1000$. If a DNA profile match is observed with s, for which $R_i = 4 \times 10^{-8}$ is accepted, while the DNA profiles of the other men are not available to the court, then we would calculate

$$P(G \mid E_\text{d}, E_\text{o}) = \frac{1}{1 + 10\,000 \times 1000 \times 4 \times 10^{-8}} = \frac{1}{1 \cdot 4} \approx 0 \cdot 71.$$

Under these assumptions, the DNA profile evidence is so strong that it still leads to s being more likely than not the culprit, but the effect of the large w_i is that the probability of G fails to be high enough for a satisfactory prosecution. If we had $w_i = 1$ for all 10 000 alternative suspects, instead of $w_i = 1000$, then the probability of guilt would be approximately 0·9996.

(c) A match with an individual who could not have committed the crime suggests that a close relative of this individual, either an identical twin if one exists or else a first-degree relative, is the true culprit.

Section 5.8, page 81

1. (a) By equation (5.3), the variance of the subpopulation allele proportions is

$$\text{Var}[\tilde{p}] = p(1-p)\theta = 0 \cdot 15 \times 0 \cdot 85 \times 0 \cdot 02 = 0 \cdot 00255,$$

and so the standard deviation is $\sqrt{0 \cdot 00255} \approx 0 \cdot 05$.

(b) Using the "plus or minus two standard deviations" rule of thumb, the 95% interval for a subpopulation proportion is $0 \cdot 15 \pm 0 \cdot 1$, which is the interval $(0 \cdot 05, 0 \cdot 25)$. Using the beta distribution with $p = 0 \cdot 15$ and $\lambda = 49$, the central 95% interval is $(0 \cdot 066, 0 \cdot 26)$. (In R use the command qbeta(c(0.025,0.975),0.15*49,0.85*49).)

2. (a) Substituting these values into (5.4) gives

$$\hat{\theta} = \frac{(11-10)^2 + (15-10)^2 + (8-10)^2 + (9-10)^2 + (12-10)^2}{5 \times 10 \times 90}$$

$$= 0 \cdot 0078.$$

SOLUTIONS TO EXERCISES

(b) If p is unknown, a natural estimate is the mean of the observed values: $(0.11 + 0.15 + 0.08 + 0.09 + 0.12)/5 = 0.11$. Substituting into (5.4), and replacing k with $k - 1$, we obtain

$$\widehat{\theta} = \frac{(11-11)^2 + (15-11)^2 + (8-11)^2 + (9-11)^2 + (12-11)^2}{4 \times 11 \times 89}$$

$$= 0.0083.$$

3. Using (5.11),

(a)
$$P(3,0) = \frac{\prod_{i=0}^{2}(i\theta + (1-\theta)p)}{(1-\theta)\prod_{i=1}^{1}(1+i\theta)}$$

$$= \frac{p(\theta + (1-\theta)p)(2\theta + (1-\theta)p)}{1+\theta},$$

(b)
$$P(2,1) = 3\frac{(1-\theta)(1-p)\prod_{i=0}^{1}(i\theta + (1-\theta)p)}{(1-\theta)\prod_{i=1}^{1}(1+i\theta)}$$

$$= 3\frac{p(1-p)(1-\theta)(\theta + (1-\theta)p)}{1+\theta}.$$

When $p = 0.25$, these formulas evaluate to

θ:	0	0.02	0.1	1
$P(3,0) \times 1000$	16	19	31	250
$P(2,1) \times 1000$	141	146	165	0

4. The allele proportion estimates are

$$\widehat{p_B} = \frac{2 \times 5 + 15 + 10}{130} = \frac{7}{26}$$

$$\widehat{p_G} = \frac{2 \times 10 + 15 + 20}{130} = \frac{11}{26}$$

$$\widehat{p_R} = \frac{2 \times 5 + 10 + 20}{130} = \frac{8}{26}$$

and so the calculations for Pearson's goodness-of-fit statistic are

Genotype	BB	GG	RR	BG	BR	GR	total
Obs. (O)	5	10	5	15	10	20	65
Exp. (E)	4.71	11.63	6.15	14.81	10.77	16.92	65
$(O-E)^2/E$	0.018	0.230	0.216	0.002	0.055	0.559	1.08

The statistic has three degrees of freedom, and since the observed value (= 1·08) is less than three, we see immediately without resort to tables that this value is not significant, and we cannot reject the hypothesis of HWE.

5. Writing a, b, c, and d for the sample counts of GR, Gr, gR, and gr haplotypes respectively, Pearson's goodness-of-fit statistic for testing the hypothesis of linkage equilibrium is (page 76):

$$nr^2 = n \times \frac{(ad-bc)^2}{(a+b)(c+d)(a+c)(b+d)}$$

$$= 43 \times \frac{(80-100)^2}{30 \times 13 \times 15 \times 28} = 0.11.$$

Again, the observed value of the statistic (= 0·11) is less than the degrees of freedom (= 1) and so we cannot reject the hypothesis of linkage equilibrium.

Section 6.6, page 109

1. (a) Consider a man u from the village, unrelated to s. At locus 1:

$$R_u^1 = 2\frac{(\theta + (1-\theta)p_A)(\theta + (1-\theta)p_B)}{(1+\theta)(1+2\theta)}$$

$$= 2\frac{(0.03 + 0.97 \times 0.06)(0.03 + 0.97 \times 0.14)}{(1+0.03)(1+0.06)} \approx 0.0268,$$

and at locus 2:

$$R_u^2 = \frac{(2\theta + (1-\theta)p_C)(3\theta + (1-\theta)p_C)}{(1+\theta)(1+2\theta)}$$

$$= 2\frac{(0.06 + 0.97 \times 0.04)(0.09 + 0.97 \times 0.04)}{(1+0.03)(1+0.06)} \approx 0.0117.$$

Using the product rule, the two-locus likelihood ratio for each of the unrelated men in the village is

$$R_u = R_u^1 \times R_u^2 \approx 0.00031.$$

Now consider h, a half-brother of s. In the notation of Section 6.2.4, for a half-brother we have $\kappa_0 = \kappa_1 = 0.5$, and

$$M_1^1 = \frac{\theta + (1-\theta)(p_A + p_B)/2}{1+\theta}$$

$$= \frac{0.03 + 0.97 \times 0.1}{1.03} \approx 0.123,$$

$$M_1^2 = \frac{2\theta + (1-\theta)p_C}{1+\theta}$$

$$= \frac{0.06 + 0.97 \times 0.04}{1.03} \approx 0.0959.$$

SOLUTIONS TO EXERCISES

By (6.5) the single-locus match probabilities for a half-brother are

$$R_h^1 = \kappa_2 + M_1^1 \kappa_1 + R_u^1 \kappa_0$$
$$\approx 0 + 0.123 \times 0.5 + 0.0268 \times 0.05 = 0.075,$$
$$R_h^2 \approx 0.0959 \times 0.5 + 0.0117 \times 0.05 = 0.054,$$

and the two-locus match probability is

$$R_h = R_h^1 \times R_h^2 \approx 0.0036.$$

Uncles have the same relationship to s as does his half-brother, and hence R_h also applies to the two uncles. Finally, using the weight-of-evidence formula (3.3), and noting that we have assumed $w_i = 1$ for all the alternative possible culprits, we have

$$P(G \mid E_d, E_o) = \frac{1}{1 + \sum w_i R_i} = \frac{1}{1 + 3R_h + 41 R_u}$$
$$\approx \frac{1}{1 + 3 \times 0.0036 + 41 \times 0.00031} = 0.977.$$

(b) Because the migrant men are from a different population, a lower value of θ is appropriate, say $\theta = 1\%$. Also, the allele proportions most appropriate for these men are those of the population from which they have come (and not, for example, those of the suspect s). Then,

$$R_m^1 = 2 \frac{(\theta + (1-\theta)p_A)(\theta + (1-\theta)p_B)}{(1+\theta)(1+2\theta)}$$
$$= 2 \frac{(0.01 + 0.99 \times 0.08)(0.01 + 0.99 \times 0.09)}{(1+0.01)(1+0.02)}$$
$$\approx 0.0172,$$
$$R_m^2 = \frac{(2\theta + (1-\theta)p_C)(3\theta + (1-\theta)p_C)}{(1+\theta)(1+2\theta)}$$
$$= 2 \frac{(0.02 + 0.99 \times 0.06)(0.03 + 0.99 \times 0.06)}{(1+0.01)(1+0.02)}$$
$$\approx 0.00689.$$

and the two-locus likelihood ratio for each of the migrant men is

$$R_m = R_m^1 \times R_m^2 \approx 0.000118.$$

The probability of guilt for s taking these 20 men into account in addition to the 44 previously considered is

$$P(G \mid E_d, E_o) = \frac{1}{1 + 3R_h + 41 R_u + 20 R_m} \approx 0.975.$$

Because of the remote possibility of coancestry with s, the 20 migrant men have a lower match probability than the men in the village, and much less than the half-brother and uncles. Taking the migrant men into account makes very little difference to the strength of the overall case against s.

2. Let r_i denote the match probability reported by the forensic scientist, which is the likelihood ratio taking only the possibility of a chance match into account, and write R_i for the likelihood ratio that allows for either a chance match or fraud. The definition of R_i is (3.1)

$$R_i = \frac{P(E \mid C = i)}{P(E \mid C = s)},$$

where E denotes the DNA evidence, and we have suppressed mention of the other evidence E_0.

We assume that the evidence E is just as likely if fraud occurs as it is if s is the true culprit, and so fraud is not an issue if $C = s$. If $C = i$, we need to consider separately the possibilities that fraud (event X) has, and has not occurred. To do this, we use a basic theorem of probability known as the Theorem of Total Probability, or the Partition Theorem (see any introductory probability textbook). In Bayesian statistics it is sometimes called "extending the conversation" because we introduce an extra variable into the probability space. The theorem gives

$$P(E \mid C=i) = P(E \mid X, C=i)P(X \mid C=i) + P(E \mid \overline{X}, C=i)P(\overline{X} \mid C=i),$$

where $\overline{X}$ denotes no error or fraud.

The probability of fraud could alter according to the identity of the true culprit: we assume this not to be the case, so that $P(X \mid C = i) = 1\%$ and $P(\overline{X} \mid C = i) = 99\%$ for all i. Moreover, if $\overline{X}$ holds, then E is just as likely as in the scenario in which the possibility of fraud is ignored. Then

$$R_i = 0 \cdot 01 + 0 \cdot 99 r_i,$$

which is $\approx 0 \cdot 01$ since $r_i = 10^{-9}$. Thus, under our assumptions, the precise value of the match probability is immaterial, beyond the fact that it is much less than the probability of fraud.

3. (a) Given that the mixed crime-scene profile and the victim's profile are both AB, the second contributor to the crime-scene DNA can have genotype AA, AB, or BB. In evaluating the probabilities of these genotypes, we use the sampling formula (5.16) and the fact that we have observed, in s and v, three A and one B alleles.

$$R_i = \frac{P(\text{AB} \mid s \equiv \text{AA}, v \equiv \text{AB}, i \text{ and } v \text{ are the sources})}{P(\text{AB} \mid s \equiv \text{AA}, v \equiv \text{AB}, s \text{ and } v \text{ are the sources})}$$
$$= P(\text{AA} \mid \text{AAAB}) + 2P(\text{AB} \mid \text{AAAB}) + P(\text{BB} \mid \text{AAAB})$$

SOLUTIONS TO EXERCISES

$$= \frac{(3\theta + (1-\theta)p_A)(4\theta + (1-\theta)p_A)}{(1+3\theta)(1+4\theta)}$$

$$+ 2\frac{(3\theta + (1-\theta)p_A)(\theta + (1-\theta)p_B)}{(1+3\theta)(1+4\theta)}$$

$$+ \frac{(\theta + (1-\theta)p_B)(2\theta + (1-\theta)p_B)}{(1+3\theta)(1+4\theta)}$$

$$= (p_A + p_B)^2 \quad \text{if } \theta = 0.$$

(b) The second contributor to the crime-scene DNA must have genotype AB, and we have previously observed one copy each of the A, B, C, and D alleles. So,

$$R_i = \frac{P(ABCD \mid s \equiv AB, v \equiv CD, i \text{ and } v \text{ are the sources})}{P(ABCD \mid s \equiv AB, v \equiv CD, s \text{ and } v \text{ are the sources})}$$

$$= 2P(AB \mid ABCD) = 2\frac{(\theta + (1-\theta)p_A)(\theta + (1-\theta)p_B)}{(1+3\theta)(1+4\theta)}$$

$$= 2p_A p_B \quad \text{if } \theta = 0.$$

(c) The required likelihood ratio is now

$$R_i = \frac{P(ABCD \mid s \equiv AB, i_1 \text{ and } i_2 \text{ are the sources})}{P(ABCD \mid s \equiv AB, s \text{ and } i_1 \text{ are the sources})}.$$

In the denominator of R_i, we must have $i_1 \equiv CD$, whereas in the numerator the genotypes of i_1 and i_2 can be AB,CD; AC,BD; or AD,BC; and each pair can occur in either order. Under our usual assumptions, all these genotype combinations have the same probability, and so we have

$$R_i = 12P(AB \mid ABCD) = 12\frac{(\theta + (1-\theta)p_A)(\theta + (1-\theta)p_B)}{(1+3\theta)(1+4\theta)}$$

$$= 12 p_A p_B \quad \text{if } \theta = 0.$$

4. (a) Because $\theta = 0$, the genotype of s_1 can be ignored. The numerator of (6.14) is unaffected, but in the denominator, AB is the only genotype possible for i_1. The probability $2p_A p_B$ cancels to leave

$$R_i = \frac{P(ABC \mid s_2 \equiv CC, i_1 \text{ and } i_2 \text{ are the sources})}{P(ABC \mid s_2 \equiv CC, s_2 \text{ and } i_1 \text{ are the sources})}$$

$$= 6p_C(p_A + p_B + p_C).$$

(b) i. The most likely genotype for the major contributor is AB, in which case R_i is just the AB genotype probability, which is $2p_A p_B$.

ii. The minor contributor can have any genotype that includes at least one C allele and no allele other than A, B, and C. Thus,

$$R_i = 2p_A p_C + 2p_B p_C + p_C^2.$$

Section 7.5, page 133

1. (a) At locus 1, ignoring mutation, the child's maternal allele is A, and hence the paternal allele is B. Since s has one C allele, the likelihood ratio when $\theta = 0$ is $2p_B$. (This exact case is not included in Table 7.1, but it is equivalent to those of rows 7 and 9 in that table.) The case of locus 2 is given in row 2 of Table 7.1: since c's paternal allele is A, and s has one A allele, the $\theta = 0$ likelihood ratio is $2p_A$. At locus 3, c's paternal allele is ambiguous; if s is the father, he must have transmitted his A allele to c, but if i is the father, he could have transmitted either A or B. The likelihood ratio is given in row 7 of Table 7.1: $2(p_A + p_B)$. Therefore,

$$\text{Overall LR} = 0.1 \times 0.2 \times 0.3 = 0.006.$$

(b) Using the formulas from Table 7.1, the overall LR is

$$\frac{0.1 + 1.9 \times 0.05}{1.05} \times \frac{0.1 + 1.9 \times 0.1}{1.05} \times \frac{0.1 + 1.9 \times 0.15}{1.05} \approx 0.0188.$$

(c) At locus 1, if s is the father, then we have observed four alleles in m and s: an A, a B, and two C (the alleles of c do not count as they are replicates of observed alleles). If i is the father, then c's allele B is distinct from that of s and makes a fifth observed allele. The likelihood ratio is twice the conditional probability of drawing a B given the observed ABCC:

$$R_i = 2P(\text{B} \mid \text{ABCC}) = 2\frac{\theta + (1-\theta)p_B}{1 + 3\theta}$$

$$= \frac{0.1 + 1.9 \times 0.05}{1.15} \approx 0.170.$$

Similarly for locus 2:

$$R_i = 2P(\text{A} \mid \text{AAAB}) = 2\frac{3\theta + (1-\theta)p_A}{1 + 3\theta}$$

$$= \frac{0.3 + 1.9 \times 0.1}{1.15} \approx 0.426.$$

For locus 3, if s is the father, we have observed AABC, whereas if i is the father we have two equally likely possibilities, A or B, and so

$$R_i = 2P(\text{A} \mid \text{AABC}) + P(\text{B} \mid \text{AABC})$$
$$= 2\frac{3\theta + (1-\theta)(p_A + p_B)}{1 + 3\theta}$$
$$= \frac{0.3 + 1.9 \times 0.15}{1.15} \approx 0.509.$$

The overall likelihood ratio is the product of these three, about 0.0368.

SOLUTIONS TO EXERCISES 171

2. (a) From Table 6.3, for a half-brother $\bar{\kappa} = 1/4$, and substituting this into (7.12) we obtain the overall likelihood ratio:

$$(0.25 + 1.5p_B) \times (0.25 + 1.5p_A) \times (0.25 + 1.5(p_A + p_B)) = 0.06175.$$

(b) The weight-of-evidence formula (3.3) applies to paternity problems in the same way as for identification. Since we are given that $w_i = 1$ for each alternative possible father, the posterior probability that s is the father is

$$P(s \text{ is father} \mid \text{profiles of } c, s, m) = \frac{1}{1 + 0.06175 + 10 \times 0.006}$$

$$= \frac{1}{1.12175} \approx 0.89.$$

We see from the calculation that the alternate possibilities that (i) the half-brother of s is the father of c, and (ii) a man unrelated to s is the father, are approximately equally likely.

3. The genotypes of c, s, and m are inconsistent with s being the father unless a mutation has occurred. The likelihood ratio for this locus when $\theta = 0$ is

$$R_i = \frac{p_A}{\mu^f_{C \to A}} = \frac{0.05}{0.0005} = 100,$$

corresponding to strong evidence that s is not the father. However, the overall likelihood ratio is $0.006 \times 100 = 0.6$, so that under our assumptions the four-locus profiles point (weakly) to s being the father rather than any particular unrelated man, despite the apparent mutation.

4. (a) The profiles of s and c do not match at any of the three loci, and so (7.14) simplifies in this case to

$$R = 0.25 + 0.5 \times R_i^p,$$

where R_i^p is the likelihood ratio for paternity when the mother is unavailable. From Table 7.2, we obtain the following expressions:

Locus	s	c	R_i^p
1	BC	AB	$4(\theta + (1-\theta)p_B)/(1+\theta)$
2	AB	AA	$2(2\theta + (1-\theta)p_A)/(1+\theta)$
3	AC	AB	$4(\theta + (1-\theta)p_A)/(1+\theta)$,

and hence we obtain, when $\theta = 0$,

$$R = (0.25 + 1/0.4) \times (0.25 + 1/0.4) \times (0.25 + 1/0.8) \approx 11,$$

and when $\theta = 0.02$,

$$R = (0.25 + 0.5 \times 1.02/(0.08 + 0.98 \times 0.2))^2$$
$$\times (0.25 + 0.5 \times 1.02/(0.08 + 0.98 \times 0.4)) \approx 5.86.$$

Thus, the data support the sibling relationship over no relation. This is unsurprising since they share an allele at each locus.

(b) Recall from (7.1) that R_i^p compares the hypotheses that s and c are unrelated (numerator) and that s and c are father – child (denominator), whereas R compares the hypotheses that s and c are siblings (numerator) and that s and c are unrelated (denominator). Thus, the required single-locus likelihood ratio, comparing siblings with father – child, is

$$R \times R_i^p = 0.5 + 0.25 \times R_i^p.$$

Evaluating over the three loci, with $\theta = 0$, we obtain

$$(0.5 + 0.05)^2 \times (0.5 + 0.1) \approx 0.18.$$

Thus, although the STR profile data support the sibling relationship over no relation, the father – child relation is supported even more. Again, this is unsurprising since, under the father – child relationship, ignoring mutation, exactly one allele will be shared ibd at each locus.

Section 8.4, page 143

1. The bound (8.4) gives

$$P(U \mid E) > 1 - 2\sum_{i \in \mathcal{P}} R_i = 1 - 2 \times 100 \times 0.01 = -1,$$

which is not very useful. However, (8.6) gives an exact formula under the assumption of independent Υ-possession, which is

$$P(U \mid E) = \frac{(1-p)^N}{1+Np}$$
$$= \frac{(1-0.01)^{100}}{1+100 \times 0.01} \approx \frac{0.366}{2} = 0.183.$$

Before the observation that s has Υ, the probability that the culprit is the unique Υ-bearer on the island is $(1 - 0.01)^{100}$ or about 37%. However this observation, together with the possibility that s is not the culprit, halves this probability to around 18%.

2. (a) The inclusion probability is

$$P(\text{Inc}) = p^2 + (1-p)^2 = 1 - 2p + 2p^2,$$

SOLUTIONS TO EXERCISES 173

where p denotes the probability of Υ possession. $P(\text{Inc}) = 1$ at both $p = 0$ and $p = 1$, and takes its optimal (i.e. minimal) value of 0.5 when $p = 0.5$. Thus, if there are only two possible outcomes for an identification test, it is best if these outcomes are equally likely. More generally, for a test with k possible outcomes, the inclusion probability has minimum value $1/k$ when all the outcomes have probability $1/k$.

(b) i. The inclusion probability of the test is

$$(0.46)^2 + (0.42)^2 + (0.09)^2 + (0.03)^2 = 0.397.$$

ii. The case-specific inclusion probabilities are equal to the individual population proportions.

The inclusion probability of the test (i) gives an average inclusion probability over many investigations and is useful for comparing this test with other possible tests but has no role in evaluating the evidence in a particular case. The case-specific inclusion probabilities (ii) do give a measure of evidential weight in a particular case, and here they are equivalent to match probabilities.

3. (a) Except for relabelling of alleles, the likelihood ratio is the same as in the example introduced on page 105, in the case $\theta = 0$:

$$R_i = \frac{12 p_B p_D (p_A + p_B + p_D)}{p_A + 2p_B + 2p_D} \approx 0.116.$$

(b) $P(\text{inclusion}) = (p_A + p_B + p_D)^2 = (0.45)^2 = 0.2025$.

(c) The exclusion probability is unaffected. For the likelihood ratio, this is essentially the same as Exercise 4(a) of Section 6.6:

$$R_i = 6 p_B (p_A + p_B + p_D) = 0.27.$$

(d) Calculating likelihood ratios becomes increasingly more complicated as the number of contributors to the sample increases, because, for example, we must consider all possible combinations of three genotypes consistent with the observed mixed profile. In contrast, the inclusion probability is unaffected by the number of contributors.

4. (a) The probability of at least one matching profile in the database is given by the binomial probability formula

$$P = 1 - (1 - p)^n.$$

Here, $n = 1000$ and $p = 10^{-6}$, and so $P \approx np = 10^{-3}$.

(b) If you said that this result was highly relevant, then either you are not paying enough attention or you have a fundamental disagreement with the author. As discussed above in Sections 3.4.5 and 6.1, we know H_0 to be false irrespective of the data, but in any case, whether or not the database was chosen randomly in the population has little relevance to the case against the specific individual, s, observed to match.

Section 9.5, page 155

1. This is a conditional statement (probability of profile *if* it came from ...), and so is probably OK.

2. Statement about the probability that Smith did not leave the blood: fallacy.

3. This is similar to the ambiguous statement discussed in the text, but it is very close to being fallacious, suggesting that there is only a 1 in 1000 chance that there is *anybody* other than Smith with the blood type.

4. Fairly standard expression of the likelihood ratio: logically OK.

5. Statement about the probability that Smith is the source: fallacy.

6. Ambiguous, but in my opinion less suggestive of fallacy than 3.

Bibliography

Aitken CGG 1995 *Statistics and the Evaluation of Evidence for Forensic Scientists*. Wiley.

Aitken CGG and Taroni F 2004 *Statistics and the Evaluation of Evidence for Forensic Scientists*, 2nd edn. Wiley. (NB only a pre-publication manuscript has been viewed by the present author)

Albanese V, Biguet NF, Kiefer H, Bayard E, Mallet J and Meloni R 2002 Quantitative effects on gene silencing by allelic variation at a tetranucleotide microsatellite. *Hum. Mol. Genet.* **10**(17), 1785–1792.

Ayres KL 2000a A two-locus forensic match probability for subdivided populations. *Genetica* **108**, 137–143.

Ayres KL 2000b Relatedness testing in subdivided populations. *Forens. Sci. Intern.* **114**, 107–115.

Ayres KL 2002 Paternal exclusion in the presence of substructure. *Forens. Sci. Intern.* **129**, 142–144.

Ayres KL and Balding DJ 1998 Measuring departures from Hardy-Weinberg: a Markov Chain Monte Carlo method for estimating the inbreeding coefficient. *Heredity* **80**, 769–778.

Ayres KL and Balding DJ 2005 Paternity index calculations when some individuals share common ancestry. *Forens. Sci. Intern.* In press.

Ayres KL, Chaseling J and Balding DJ 2002 Implications for DNA identification arising from an analysis of Australian forensic databases. *Forens. Sci. Intern.* **129**, 90–98.

Ayres KL and Overall ADJ 1999 Allowing for within-subpopulation inbreeding in forensic match probabilities. *Forens. Sci. Intern.* **103**, 207–216.

Ayres KL and Powley WM 2004 Calculating the exclusion probability and paternity index for X-chromosomal loci in the presence of substructure. *Forens. Sci. Intern.* In press

Balding DJ 1995 Estimating products in forensic identification using DNA profiles. *J. Am. Statist. Assoc.* **90**, 839–844.

Balding DJ 1997 Errors and misunderstandings in the second NRC report. *Jurimetrics* **37**, 469–476.

Balding DJ 1999 When can a DNA profile be regarded as unique? *Sci. Just.* **39**, 257–260.

Balding DJ 2000 Interpreting DNA evidence: can probability theory help? In *Statistical Science in the Courtroom* (ed. Gastwirth J), pp. 51–70. Springer-Verlag.

Balding DJ 2002 The DNA database controversy. *Biometrics* **58**, 241–244.

Balding DJ 2003 Likelihood-based inference for genetic correlation coefficients. *Theor. Pop. Biol.* **63**(3), 221–230.

Balding DJ and Donnelly P 1994 The prosecutor's fallacy and DNA evidence. *Criminal Law Rev.* **October**, 711–721.

Balding DJ and Donnelly P 1995a Inference in forensic identification. *J. R. Statist. Soc.* A **158**, 21–53.

Balding DJ and Donnelly P 1995b Inferring identity from DNA profile evidence. *Proc. Natl. Acad. Sci. U.S.A.* **92**, 11 741–11 745.

Balding DJ and Donnelly P 1996 Evaluating DNA profile evidence when the suspect is identified through a database search. *J. Forens. Sci.* **41**, 603–607.

Balding DJ, Greenhalgh M and Nichols RA 1996 Population genetics of STR loci in Caucasians. *Int. J. Legal Med.* **108**, 300–305.

Balding DJ and Nichols RA 1994 DNA profile match probability calculation: how to allow for population stratification, relatedness, database selection and single bands. *Forens. Sci. Intern.* **64**, 125–140.

Balding DJ and Nichols RA 1995 A method for quantifying differentiation between populations at multi-allelic loci and its implications for investigating identity and paternity. *Genetica* **96**, 3–12.

Balding DJ and Nichols RA 1997 Significant genetic correlations among Caucasians at forensic DNA loci. *Heredity* **78**, 583–589.

Bamshad MJ, Wooding S, Watkins WS, Ostler CT, Batzer MA and Jorde LB 2003 Human population genetic structure and inference of group membership. *Am. J. Hum. Genet.* **72**, 578–589.

Bernardo JM and Smith AFM 1994 *Bayesian Theory*. Wiley.

Birky WCJ 2001 The inheritance of genes in mitochondria and chloroplasts: laws, mechanisms, and models. *Annu. Rev. Genet.* **35**, 125–148.

Brenner CH 1998 Difficulties in the estimation of ethnic affiliation. *Am. J. Hum. Genet.* **62**(6), 1558–1560.

Brenner CH 2003 Forensic Genetics: Mathematics, *Encyclopedia of the Human Genome*, pp. 513-519. Macmillan, London.

Brenner CH and Weir BS 2003 Issues and strategies in the DNA identification of World Trade Center victims. *Theor. Pop. Biol.* **63**(3), 173–178 .

Buckleton JS, Evett IW and Weir BS 1998 Setting bounds for the likelihood ratio when multiple hypotheses are postulated. *Sci. Just.* **38**(1), 23–26.

Buckleton JS, Triggs CM and Walsh SJ 2004 *Forensic DNA Evidence Interpretation*. CRC Press, Boca Raton, FL. (NB only a pre-publication manuscript has been viewed by the present author)

Butler JM 2001 *Forensic DNA Typing: Biology and Technology Behind STR Markers*. Academic Press, ISBN 01 21 47 951X.

Butler JM, Schoske R, Vallone PM, Redman JW and Kline MC 2003 Allele frequencies for 15 autosomal STR loci on U.S. Caucasian, African American and Hispanic populations. *J. Forens. Sci.* **48**, 1–4.

Calabrese P and Durrett R 2003 Dinucleotide repeats in the Drosophila and human genomes have complex, length-dependent mutation processes. *Mol. Biol. Evol.* **20**(5), 715–725.

Chakraborty R, Srinivasan MR and Daiger SP 1993 Evaluation of standard errors and confidence intervals of estimated multilocus genotype probabilities, and their implications in DNA forensics. *Am. J. Hum. Genet.* **52**(1), 60–70.

BIBLIOGRAPHY

Chakravarti A and Li CC 1984 Estimating the prior probability of paternity from the results of exclusion tests. *Forens. Sci. Intern.* **24**(2), 143–147.

Clayton TM, Whitaker JP, Sparkes R and Gill P 1998 Analysis and interpretation of mixed forensic stains using DNA STR profiling. *Forens. Sci. Intern.* **91**(1), 55–70.

Cook R, Evett IW, Jackson G, Jones PJ and Lambert JA 1998 A hierarchy of propositions: deciding which level to address in casework. *Sci. Just.* **38**, 231–240.

Cowell RG 2003 FINEX: a probabilistic expert system for forensic identification. *Forens. Sci. Intern.* **134**, 196–206.

Curran JM, Buckleton JS, Triggs CM and Weir BS 2002 Assessing uncertainty in DNA evidence caused by sampling effects. *Sci. Just.* **42**(1), 29–37.

Curran JM, Triggs CM, Buckleton J and Weir BS 1999 Interpreting DNA mixtures in structured populations. *J. Forens. Sci.* **44**(5), 987–995.

Dawid AP 2001 Comment on Stockmarr's Likelihood ratios for evaluating DNA evidence when the suspect is found through a database search. *Biometrics* **57**, 976–978.

Dawid AP 2004 Which likelihood ratio (Comment on "Why the effect of prior odds should accompany the likelihood ratio when reporting DNA evidence" by Ronald Meester and Marjan Sjerps). *Law, Probab. Risk* **3**, 65–71.

Dawid AP and Mortera J 1996 Coherent analysis of forensic identification evidence. *J. R. Statist. Soc.* **B58**, 425–444.

Dawid AP and Mortera J 1998 Forensic identification with imperfect evidence. *Biometrika* **85**, 835–849.

Dawid AP, Mortera J and Pascali VL 1996 Non-fatherhood or mutation? A probabilistic approach to parental exclusion in paternity testing. *Forens. Sci. Intern.* **124**(1), 55–61.

Dawid AP, Mortera J, Pascali VL and van Boxel DW 2002 Probabilistic expert systems for forensic inference from genetic markers. *Scand. J. Statist.* **29**, 577-595.

Donnelly P and Friedman RD 1999 DNA database searches and the legal consumption of scientific evidence. *Mich. Law Rev.* **97**(4), 931–984.

Egeland T, Dalen I and Mostad PF 2003 Estimating the number of contributors to a DNA profile. *Int. J. Legal Med.* **117**(5), 271–275.

Egeland T, Mostad PF, Mevag B and Stenersen M 2000 Beyond traditional paternity and identification cases: selecting the most probable pedigree. *Forens. Sci. Intern.* **110**(1), 47–59.

Eggleston R 1983 *Evidence, Proof and Probability*, 2nd edn. Wiedenfield and Nicholson, London.

Evett IW 1995 Avoiding the transposed conditional. *Sci. Just.* **35**, 127–131.

Evett IW, Buffery C, Willott G and Stoney D 1991 A Guide to interpreting single locus profiles of DNA mixtures in forensic cases. *J. Forens. Sci. Soc.* **31**(1), 41–47.

Evett IW, Gill PD, Jackson G, Whitaker J and Champod C 2002 Interpreting small quantities of DNA: the hierarchy of propositions and the use of Bayesian networks. *J. Forens. Sci.* **47**(3), 520–530.

Evett IW, Gill PD and Lambert JA 1998 Taking account of peak areas when interpreting DNA mixtures. *J. Forens. Sci.* **43**(1), 62–69.

Evett IW, Jackson G and Lambert JA 2000 More in the hierarchy of propositions: exploring the distinctions between explanations and propositions. *Sci. Just.* **40**, 3–10.

Evett IW and Weir BS 1998 *Interpreting DNA Evidence*. Sinauer, Sunderland, MA.

Finkelstein MO and Fairley WB 1970 A Bayesian approach to identification evidence. *Harv. Law Rev.* **83**(3), 489–517.

Finkelstein MO and Fairley WB 1971 A comment on "Trial by mathematics". *Harv. Law Rev.* **84**, 1800–1810.

Foreman LA, Champod C, Evett IW, Lambert JA and Pope S 2003 Interpreting DNA evidence: a review. *Intern. Statist. Rev.* **71**(3), 473-495.

Frudakis T, Venkateswarlu K, Thomas MJ, Gaskin Z, Ginjupalli S, Gunturi S, Ponnuswamy V, Natarajan S and Nachimuthu PK 2003 A classifier for the SNP-based inference of ancestry. *J. Forens. Sci.* **48**(4), 771–782.

Fung WK 2003 User-friendly programs for easy calculations in paternity testing and kinship determinations. *Forens. Sci. Intern.* **136**, 22–34.

Fung WK, Carracedo A and Hu YQ 2004 Testing for kinship in a subdivided population. *Forens. Sci. Intern.* **135**(2), 105–109.

Fung WK and Hu YQ 2000 Interpreting forensic DNA mixtures: allowing for uncertainty in population substructure and dependence. *J. R. Statist. Soc.* A **163**, 241–254.

Fung WK and Hu YQ 2002 The statistical evaluation of DNA mixtures with contributors from different ethnic groups. *Int. J. Legal Med.* **116**(2), 79–86.

Fung WK and Hu YQ 2004 Interpreting DNA mixtures with related contributors in subdivided populations. *Scand. J. Statist* **31**(1), 115–130.

Gill P 2001a Application of low copy number DNA profiling. *Croat. Med. J.* **42**, 229–232.

Gill P 2001b An assessment of the suitability of single nucleotide polymorphisms (SNPs) for forensic purposes. *Int. J. Legal Med.* **114**, 204–210.

Gill P 2002 The role of short tandem repeat (STR) DNA in forensic casework in the UK–past, present and future. *Biotechniques* **22**, 366–385.

Gill P, Ivanov PL, Kimpton C, Piercy R, Benson N, Tully G, Evett I, Hagelberg R and Sullivan K 1994 Identification of the remains of the Romanov family by DNA analysis. *Nat. Genet.* **6**(2), 130–135.

Gill P, Sparkes R, Pinchin R, Clayton T, Whitaker J and Buckleton J 1998 Interpreting simple STR mixtures using allele peak areas. *Forens. Sci. Intern.* **91**(1), 41–53.

Gill P, Werrett DJ, Budowle B and Guerrieri R 2004 An assessment of whether SNPs will replace STRs in national DNA databases–Joint considerations of the DNA working group of the European Network of Forensic Science Institutes (ENFSI) and the Scientific Working Group on DNA Analysis Methods (SWGDAM). *Sci. Just.* **44**(1), 51–53.

Good IJ 1991 Weight of evidence and the Bayesian likelihood ratio. Chapter 3. In *The Use of Statistics in Forensic Science* (eds Aitken CGG and Stoney DA). Ellis Horwood.

Good IJ 1995 When batterer turns murderer. *Nature* **375**, 541.

Guo S-W and Thompson EA 1992 Performing the exact test of Hardy-Weinberg proportion for multiple alleles. *Biometrics* **48**, 361–372.

Hu YQ and Fung WK 2003 Evaluating forensic DNA mixtures with contributors of different structured ethnic origins: a computer software. *Intern. J. Legal Med.* **117**(4), 248–249.

Jobling MA, Hurles ME and Tyler-Smith C 2004 *Human Evolutionary Genetics: Origins, Peoples, and Disease.* Garland Science.

Jobling MA, Pandya A and Tyler-Smith C 1996 The Y chromosome in forensic analysis and paternity testing. *Intern. J. Legal Med.* **110**(3), 118–124.

Johnson P and Williams R 2004 Post-conviction DNA testing: the UK's first 'exoneration' case? *Sci. Just.* **44**(2), 77–82.

Kaye DH 1989 The probability of an ultimate issue: the strange cases of paternity testing. *Iowa Law Rev.* **75**(1), 75–109.

Kaye DH 2003 Questioning a courtroom proof of the uniqueness of fingerprints. *Intern. Statist. Rev.* **71**, 521–534.

Kaye DH and Sensabaugh GF 2000 Reference guide on DNA evidence, *Reference Manual on Scientific Evidence*, 2nd edn. Federal Judicial Center, www.fjc.gov

Kaye DH and Sensabaugh GF 2002 DNA typing. In *Modern Scientific Evidence: The Law and Science of Expert Testimony*, 2nd edn (eds Faigman, Kaye and Saks). West Publishing Company, St Paul, MN.

Kaye DH and Smith ME 2003 DNA identification databases: legality, legitimacy, and the case for population-wide coverage. *Wis. Law Rev.* **2003**(3), 413–459.

Kimura M 1977 Preponderance of synonymous changes as evidence for the neutral theory of molecular evolution. *Nature* **267**, 275–276.

Koehler JJ 1993a DNA matches and statistics: important questions, surprising answers. *Judicature* **76**(5), 222–229.

Koehler JJ 1993b Error and exaggeration in the presentation of DNA evidence at trial. *Jurimetrics* **34**, 21–35.

Koehler JJ 1996 On conveying the probative value of DNA evidence: frequencies, likelihood ratios, and error rates. *Univ. Colo. Law Rev.* **67**(4), 859–886.

Koehler JJ 2001 When are people persuaded by DNA match statistics? *Law Hum. Behav.* **25**(5), 493–513.

Lai YL and Sun FZ 2003 The relationship between microsatellite slippage mutation rate and the number of repeat units. *Mol. Biol. Evol.* **20**(12), 2123–2131.

Lambert JA and Evett IW 1998 The impact of recent judgements on the presentation of DNA evidence. *Sci. Just.* **38**(4), 266–270.

Lauritzen SL and Mortera J 2002 Bounding the number of contributors to mixed DNA stains. *Forens. Sci. Intern.* **130**(2-3), 125–126.

Lauritzen SL and Sheehan NA 2003 Graphical models for genetic analyses. *Statist. Sci.* **18**(4), 489–514.

Lewontin RC 1972 The apportionment of human diversity. *Evol. Biol.* **6**, 381–398.

Lowe AL, Urquart A, Foreman LA and Evett IW 2001 Inferring ethnic origin by means of an STR profile. *Forens. Sci. Intern.* **119**, 17–22.

Li CC and Chakravarti A 1988 An expository review of 2 methods of calculating the paternity probability. *Am. J. Hum. Genet.* **43**, 197–205.

Maiste PJ and Weir BS 1995 A comparison of tests for independence in the FBI RFLP databases. *Genetica* **96**, 125–138.

Marchini J, Cardon LR, Phillips MS and Donnelly P 2004 The effects of human population structure on large genetic association studies. *Nat. Genet.* **36**, 512–517.

Mayor L and Balding DJ 2005 Discrimination of half-siblings in structured populations when maternal genotypes are unknown. *Forens. Sci. Intern.* In press.

Meester R and Sjerps M 2003 The evidential value in the DNA database search controversy and the two-stain problem. *Biometrics* **59**, 727–732.

Meester R and Sjerps M 2004 Why the effect of prior odds should accompany the likelihood ratio when reporting DNA evidence. *Law, Probab. Risk* **3**, 51–62. (See also the discussion and response, pp 63–86.)

Mortera J, Dawid AP and Lauritzen SL 2003 Probabilistic expert systems for DNA mixture profiling. *Theor. Pop. Biol.* **63**(3), 191–205.

National Research Council 1992 *DNA Technology in Forensic Science*. National Academy Press, Washington, DC.

National Research Council 1996 *The Evaluation of Forensic DNA Evidence.* National Academy Press, Washington, DC.

Parson W, Brandstatter A, Alonso A, Brandt N, Brinkmann B, Carracedo A, Corach D, Froment O, Furac I, Grzybowski T, Hedberg K, Keyser-Tracqui C, Kupiec T, Lutz-Bonengel S, Mevag B, Ploski R, Schmitter H, Schneider P, Syndercombe-Court D, Sorensen E, Thew H, Tully G and Scheithauer R 2004 The EDNAP mitochondrial DNA population database (EMPOP) collaborative exercises: organisation, results and perspectives. *Forens. Sci. Intern.* **139**(2-3), 215–226.

Pueschel J 2001 *The Application of Bayesian Hierarchical Models to DNA Profiling Data.* Ph.D. Thesis, Department of Statistical Science, University College, London.

Rannala B and Mountain JL 1997 Detecting immigration by using multilocus genotypes. *Proc. Natl. Acad. Sci. U.S.A.* **17**, 197–201.

Robertson B and Vignaux GA 1995 *Interpreting Evidence.* Wiley.

Roeder K 1994 DNA fingerprinting: a review of the controversy. *Statist. Sci.* **9**, 222-278.

Roewer L, Kayser M, de Knijff P, Anslinger K, Betz A, Caglia A, Corach D, Furedi S, Henke L, Hidding M, Kargel HJ, Lessig R, Nagy M, Pascali VL, Parson W, Rolf B, Schmitt C, Szibor R, Teifel-Greding J and Krawczak M 2000 A new method for the evaluation of matches in non-recombining genomes: application to Y-chromosomal short tandem repeat (STR) haplotypes in European males. *Forens. Sci. Intern.* **114**, 31–43.

Romualdi C, Balding DJ, Nasidze IS, Risch G, Robichaux M, Sherry S, Stoneking M, Batzer MA and Barbujani G 2002 Patterns of human diversity, within and among continents, inferred from biallelic DNA polymorphisms. *Genome Res.*, **12**, 602–612.

Rosenberg NA, Li LM, Ward R and Pritchard JK 2003 Informativeness of genetic markers for inference of ancestry. *Am. J. Hum. Genet.* **73**, 1402–1422.

Rosenberg NA, Pritchard JK, Weber JL, Cann HM, Kidd KK, Zhivotovsky LA and Feldman MW 2002 Genetic structure of human populations. *Science* **298**(5602), 2381–2385.

Rudin N and Inman K 2002 *An Introduction to Forensic DNA Analysis*, 2nd edn. CRC Press, Boca Raton, FL.

Schaid DJ 2004 Linkage disequilibrium testing when linkage phase is unknown. *Genetics* **166**, 505–512.

Shriver MD, Smith MW and Jin L 1998 Difficulties in the estimation of ethnic affiliation–Reply. *Am. J. Hum. Genet.* **62**(6), 1560–1561.

Shriver MD, Smith MW, Jin L, Marcini A, Akey JM, Deka R and Ferrell RE 1997 Ethnic-affiliation estimation by use of population-specific DNA markers. *Am. J. Hum. Genet.* **60**(4), 957–964.

Sjerps M and Kloosterman AD 1999 On the consequences of DNA mismatches for close relatives of an excluded suspect. *Intern. J. Legal Med.* **112**, 176–180.

Stockmarr A 1999 Likelihood ratios for evaluating DNA evidence when the suspect is found through a database search. *Biometrics*, **55**, 671–677.

Subramanian S, Mishra RK and Singh L 2003 Genome-wide analysis of Bkm sequences (GATA repeats): predominant association with sex chromosomes and potential role in higher order chromatin organization and function. *Bioinformatics* **19**(6), 681–685.

Szibor R, Krawczak M, Hering S, Edelmann J, Kuhlisch E and Krause D 2003 Use of X-linked markers for forensic purposes. *Intern. J. Legal Med.* **117**, 67–74.

Thompson WC 1989 Are jurors competent to evaluate statistical evidence? *Law Contemp. Probl.* **52**(4), 9–41.

Thompson WC and Schumann EL 1987 Interpretation of statistical evidence in criminal trials: the prosecutor's fallacy and the defense attorney's fallacy. *Law Hum. Behav.* **11**(3), 167–187.

Thompson WC, Taroni F and Aitken CGG 2003 How the probability of a false positive affects the value of DNA evidence. *J. Forens. Sci.* **48**(1), 47–54.

Templeton AR 1999 Human races: a genetic and evolutionary perspective. *Am. J. Anthropol.* **100**, 632–650.

Torres Y, Flores I, Prieto V, Lopez-Soto M, Farfan MJ, Carracedo A and Sanz P 2003 DNA mixtures in forensic casework: a 4-year retrospective study. *Forens. Sci. Intern.* **134**(2-3), 180–186.

Tribe LH 1971 Trial by mathematics: precision and ritual in the legal process. *Harv. Law Rev.* **84**(6), 1329–1393.

Tully G, Bär W, Brinkmann B, Carracedo A, Gill P, Morling N, Parson W and Schneider P 2001 Considerations by the European DNA profiling (EDNAP) group on the working practices, nomenclature and interpretation of mitochondrial DNA profiles. *Forens. Sci. Intern.* **124**(1), 83–91.

Vallone PM, Just RS, Coble MD, Butler JM and Parsons TJ 2004 A multiplex allele-specific primer extension assay for forensically informative SNPs distributed throughout the mitochondrial genome. *Intern. J. Legal Med.* **118**(3), 147–157.

Weir BS 1994 The effects of inbreeding on forensic calculations. *Annu. Rev. Genet.* **28**, 597–621.

Weir BS 1996 *Genetic Data Analysis II*. Sinauer, Sunderland, MA.

Weir BS 2003 Forensics. In *Handbook of Statistical Genetics*, 2nd edn (eds Balding DJ, Bishop M and Cannings C), pp. 830–852. Wiley.

Weir BS and Hill WG 2002 Estimating F-Statistics. *Annu. Rev. Genet.* **36**, 721–50.

Whittaker JC, Harbord RM, Boxall N, Mackay I, Dawson G and Sibly RM 2003 Likelihood-based estimation of microsatellite mutation rates. *Genetics* **164**(2), 781-787.

Wilson IJ, Weale M and Balding DJ 2003 Inferences from DNA data: population histories, evolutionary processes, and forensic match probabilities. *J. R. Statist. Soc.* A **166**(2), 155-187.

Wiuf C 2001 Recombination in human mitochondrial DNA? *Genetics* **159**, 749–756.

Wright S 1951 The genetical structure of populations. *Ann. Eugen.* **15**, 323–354.

Xu X, Peng M, Fang Z and Xu X 2000 The direction of microsatellite mutations is dependent upon allele length. *Nat. Genet.* **24**, 396-399.

Zarrabeitia MT, Riancho JA, Lareu MV, Leyva-Cobiaan F and Carracedo A 2003 Significance of micro-geographical population structure in forensic cases: a Bayesian exploration. *Intern. J. Legal Med.* **117**(5), 302–305.

Zaykin D, Zhitovsky L and Weir BS 1995 Exact tests for association between alleles at arbitrary numbers of loci. *Genetica* **96**, 169–178.

Zerjal T, Xue YL, Bertorelle G, Wells RS, Bao WD, Zhu SL, Qamar R, Ayub Q, Mohyuddin A, Fu SB, Li P, Yuldasheva N, Ruzibakiev R, Xu JJ, Shu QF, Du RF, Yang HM, Hurles ME, Robinson E, Gerelsaikhan T, Dashnyam B, Mehdi SQ and Tyler-Smith C 2003 The genetic legacy of the Mongols. *Am. J. Hum. Genet.* **72**(3), 717–721.

Index

admixture 62, 76, 90, 95
alibi 30, 31, 33, 41, 86, 149

database
    intelligence 35, 36, 85, 94
    population 35, 62, 63, 75, 86,
        87, 93–95, 99, 101, 114,
        136
distribution
    beta-binomial 64–67, 69
    beta/Dirichlet 63–64, 68, 69,
        77, 101
    binomial 4, 56, 173
    normal 93

electropherogram (EPG) 44, 46,
    47, 101–103, 108–110
error
    contamination 17, 49, 98
    typing/handling 2, 3, 9,
        15–17, 25, 34–35, 40, 70,
        71, 75, 98, 139
estimation
    conservative 90, 94, 96, 100,
        140
    of f 73–74
    of p 63, 87, 100
    of θ 79–81, 96, 97, 100
    of mixture proportion
        108–109
    "plug-in" 87, 93–94, 96, 97

ethnic (or racial) group 15, 19, 30,
    96, 97, 115, 131

fallacy
    defendant's 147–148
    prosecutor's 146–149, 151
    uniqueness 148
    weight-of-evidence 146, 147

grouping alternative suspects 39

heteroplasmy 51, 99–101
hypothesis tests 4, 74, 93,
    141–143, 166

inbreeding 62, 70, 73–75, 88, 90,
    91, 135

linkage 53, 75–76, 117–118
    (dis)equilibrium (LE/LD)
        75–76, 90, 100, 132,
        166

mitochondrial DNA (mtDNA) 43,
    50–51, 54, 59, 61,
    99–101
mixture profiles 48, 49, 101–104,
    110, 131, 136, 140
mutation 48, 59–60, 69, 122–123,
    125–126

order of evidence 27–28, 30, 152

R (statistical software)   64, 72, 73, 164

population structure   60, 70, 100, 101

relatedness coefficients   91, 119, 126–127, 171

selection   52, 60–62, 70, 75, 76, 86, 100

single-nucleotide polymorphism (SNP)   4, 51, 52, 53–54, 69, 104, 133

victim   31, 33, 41, 83, 104, 106, 110, 127, 149, 161, 168

Y chromosome   51–52, 60, 101

Statistics in Practice

Human and Biological Sciences

Brown and Prescott – Applied Mixed Models in Medicine
Ellenberg, Fleming and DeMets – Data Monitoring Committees in Clinical Trials: A Practical Perspective
Lawson, Browne and Vidal Rodeiro – Disease Mapping with WinBUGS and MLwiN
Lui – Statistical Estimation of Epidemiological Risk
*Marubini and Valsecchi – Analysing Survival Data from Clinical Trials and Observation Studies
Parmigiani – Modeling in Medical Decision Making: A Bayesian Approach
Senn – Cross-over Trials in Clinical Research, Second Edition
Senn – Statistical Issues in Drug Development
Spiegelhalter, Abrams and Myles – Bayesian Approaches to Clinical Trials and Health-Care Evaluation
Whitehead – Design and Analysis of Sequential Clinical Trials, Revised Second Edition
Whitehead – Meta-Analysis of Controlled Clinical Trials

Earth and Environmental Sciences

Buck, Cavanagh and Litton – Bayesian Approach to Interpreting Archaeological Data
Glasbey and Horgan – Image Analysis in the Biological Sciences
Helsel, Nondetects and Data Analysis: Statistics for Censored Environmental Data
Webster and Oliver – Geostatistics for Environmental Scientists

Industry, Commerce and Finance

Aitken and Taroni – Statistics and the Evaluation of Evidence for Forensic Scientists, Second Edition
Balding – Weight-of-evidence for Forensic DNA Profiles
Lehtonen and Pahkinen – Practical Methods for Design and Analysis of Complex Surveys, Second Edition
Ohser and Mücklich – Statistical Analysis of Microstructures in Materials Science

*Now available in paperback

Printed and bound by CPI Group (UK) Ltd, Croydon, CR0 4YY
09/06/2025

14685967-0001